SOME INVENTORY MODELS APPROACH TO DECISION MAKING IN FUZZY ENVIRONMENT

First Edition

Wasim Akram Mandal, Ph.D.
Beldanga D.H.Sr Madrasah

Sahidul Islam, Ph.D.
University of Kalyani

WASIM AKRAM MANDAL is an assistant teacher of Beldanga D. H. Sr. Madrasah, Beldanga, Murshidabad, West Bengal, India. He has completed his PhD under the guidance of Dr. Sahidul Islam.

SAHIDUL ISLAM is an assistant professor of University of Kalyani, Kalyani, Nadia, West Bengal, India.

ISBN-13:978-1973846604
ISBN-10:1973846608

If you liked this book, please rate it on Amazon.com
Should you have any question or suggestion, fell free to contact us at: **wasim0018@gmail.com**

"DO NOT WORRY ABOUT YOUR PROBLEMS WITH MATHEMATICS, I ASSURE YOU MINE ARE FAR GREATER."

Albert Einstein

DEDICATION

To my family.......

PREFACE

Mathematics, the king of all sciences, remains and will remain as a subject with great charm having an intrinsic value and beauty of its own. It plays very important role in sciences, engineering and other subjects as well. So, mathematical knowledge is essential for the growth of science and technology, and for any individual to shine well in the field of one's choice. In addition, a rigorous mathematical training gives one not only the knowledge of mathematics but also a disciplined thought process, an ability to analyze complicated problems. We need the power and prowess of mathematics to face and solve the ever increasing complex problems that we encounter in our life. Furthermore, mathematics is a supremely creative force and not just a problem solving tool. The learners will realize this fact to their immense satisfaction and advantage as they learn more and more of mathematics. Besides, a good mathematical training is very much essential to create a good work force for posterity. The rudiments of mathematics attained at the school level form the basis of higher studies in the field of mathematics and other sciences. Besides learning the basics of mathematics, it is also important to learn how to apply them in solving problems.

The basic objective of an inventory model is to reduce total average cost, i.e., maximum profit and also ensuring that the production process does not suffer at the same time. In this book, we have developed some inventory models for finding total annual average cost and optimal ordering quantity in crisp and fuzzy environment. In the past researchers assumed the parameters involved in an inventory model such as the demand, holding cost, se-up cost, deteriorating cost, etc. as a crisp values or random variables. But in reality some uncertainty occurs. The demand and cost of the items change from day to day. Again, calculating these variables by the probability distribution is very hard due to lack of historical data. The cost parameters generally estimated based on previous experiment and managerial judgment. Therefore, in real life situation fuzzy set theory is more realistic than crisp set theory or traditional probability theory. Here we have used triangular fuzzy number (TFN) also used fuzzy grade-mean integration representation (GMIR) method. Optimal solution (OS) is obtained here by applying non-linear programming techniques (GP, SGP, Khun –Tucker necessary conditions) for a single item inventory model. But it is observed that the geometric programming (GP) technique minimizes the total annual average cost more than any usual non-linear technique. It is also analyzed that when fuzzification of the given inventory model is done the total average cost is minimized by satisfying the constraints under necessary conditions.

Finally, we wish to thank the staff of the **Kindle Amazon** for their unfailing cooperation. We invite constructive suggestions from any reader of this book.

Kolkata
July, 2017

Wasim Akram Mandal
Sahidul Islam

ABSTRACT

Inventory management (IM) or inventory control is an important field for real world applications and research purpose. The basic objective of an inventory management (IM) is to minimize the investment in an inventory at minimum level to maximize profitability also keeping inventory at sufficiently high level to perform production and sales activities smoothly. In this thesis, some inventory models are developed for finding total average cost (TAC) and optimal ordering quantity in the system for single item in crisp and fuzzy environment. Optimal solution (OS) is obtained by applying several non-linear programming techniques (GP, SGP, FGP, Khun –Tucker necessary conditions).This thesis is in the field of inventory management (IM) contains eleven chapters.

In the first chapter, we have discussed some basic concepts of operations research (OR), inventory management (IM) and fuzzy set theory. **In the second chapter,** we have analyzed some basic techniques and methodology which is used for solving inventory models in crisp and fuzzy environment. **In the third chapter,** an economic order quantity (EOQ) model with cost of interest, time dependent holding cost, without shortages is formulated and solved. Here the inventory model is solved by geometric programming (GP), non-linear programming (NLP), fuzzy geometric programming (FGP) and fuzzy non-linear programming (FNLP) technique respectively. **In the fourth chapter,** we have developed a fuzzy inventory model for deterioration items with constant demand. Shortages are allowed under fully backlogged. In this model a special condition that is, the demand falls to zero in a time interval ($t_0 \leq t \leq t_e$) for an unexpected situation (flood, strike, earthquake, etc.) is considered. **In the fifth chapter,** an inventory model with unit production cost, time dependent holding cost, without shortages is formulated and solved. Geometric programming (GP) technique is used here for solving inventory model. Also nearest interval approximation (NIA) method is used here to convert a triangular fuzzy number (TFN) to an interval number then transform this interval number to a parametric interval-valued functional form and solve the parametric problem by geometric programming (GP) technique. **In the sixth chapter,** an economic order quantity (EOQ) model with unit production cost, time dependent holding cost, without shortages is formulated and solved. The problem is solved using fuzzy max-min geometric programming (GP) technique and fuzzy parametric geometric programming (GP) technique respectively. Sensitivity analysis is also presented here. **In the seventh chapter,** we have analyzed fuzzy inventory model for deterioration items with time dependent demand. Shortages are allowed under fully backlogged. In fuzzy environment we have considered all required parameters to be triangular fuzzy numbers. **In the eighth chapter,** we have proposed a single item inventory model with constant demand, without shortages in a fuzzy environment. In this model, we have developed the concepts of possibility theory and possibilistic moment generating function and some statistical concept as mean, variance, standard deviation on this economic order quantity (EOQ) model. **In the ninth chapter,** we have proposed an inventory model with shortages under fully backlogging and constant demand in a crisp and fuzzy environment. Here a new idea, that is Bell shaped fuzzy membership function (MF) is developed, also developed the concepts of possibility theory and possibilistic moment generating function. Three type of possibilistic mean values as lower possibilistic mean$(E_L(A))$, upper possibilistic mean $(E_R(A))$ and crisp possibilistic mean (E(A)) of total average cost (TAC) function is developed here. **In the tenth chapter,** a fuzzy economic order quantity (EOQ) model with shortages under fully backlogging and constant demand is formulated and solved. Here we have derived a new idea that is, fuzzy modified signomial geometric programming (FMSGP) technique. The last chapter, i.e., **eleventh chapter,** is Conclusions & future work.

TABLE OF CONTENTS

GLOSSARY OF TERMS

AM	Arithmetic Mean
CGP	Constrained Geometric Programming
CMGP	Constrained Modified Geometric Programming
CMSGP	Constrained Modified Signomial Geometric Programming
CSGP	Constrained Signomial Geometric Programming
DD	Degree of Difficulty
DM	Degree of Membership
DP	Dual Programming
EOQ	Economic Order Quantity
EPQ	Economic Production Quantity
etc.	Etcetera
FGMI	Fuzzy Grade-Mean Integration Method
FGP	Fuzzy Geometric Programming
FMODM	Fuzzy Multi-Objective Decision Making
FMSGP	Fuzzy Modified Signomial Geometric Programming
FNLP	Fuzzy Non-Linear Programming
FPGP	Fuzzy Parametric Geometric Programming
FSGP	Fuzzy Signomial Geometric Programming
GFN	Generalized Fuzzy Number
GM	Geometric Mean
GP	Geometric Programming
GTrFN	Generalized Trapezoidal Fuzzy Number
I.e.,/i.e.,	Therefore
IM	Inventory Management
LOG/log	Logarithm
LPP	Linear Programming Problem
LS	Least Square
MAX/Max	Maximum
MF	Membership Function
MIN/Min	Minimum

MODM	Multi-Objective Decision Making
MSGP	Modified Signomial Geometric Programming
NIA	Nearest Interval Approximation
NLP	Non-Linear Programming
OM	Operations Management
OOR	Out-Of-Stock
OR	Operations Research
OS	Optimal Solution
PfFN	Parabolic Flat Fuzzy Number
PFN	Pentagonal Fuzzy Number
PGP	Parametric Geometric Programming
PrFN	Parabolic Fuzzy Number
PSGP	Parametric Signomial Geometric Programming
PSMGP	Parametric Signomial Modified Geometric Programming
SD	Standard Deviation
SGP	Signomial Geometric Programming
TAC/Tac	Total Average Cost
TFN	Triangular Fuzzy Number
TrFN	Trapezoidal Fuzzy Number
UGP	Unconstrained Geometric Programming
UMGP	Unconstrained Modified Geometric Programming
UMSGP	Unconstrained Modified Signomial Geometric Programming
USGP	Unconstrained Signomial Geometric Programming
VAR/Var	Variance

LIST OF FIGURES

LIST OF TABLES

"Watch your thoughts; they become words.
Watch your words; they become actions.
Watch your actions; they become habits.
Watch your habits; they become character.
Watch your character; it becomes your destiny"

Lao-Tze

Chapter 1

Introduction

This chapter briefly describes the background of operations research (OR), inventory models and fuzzy sets theory for this research work. Literature review and organization of thesis are also presented here. Operations research (OR) or operations management (OM) is one of the important parts of applied mathematics, mainly used in optimization. Operations research (OR) is engaged with coordinating and controlling the operations or activities within the organization. It is one of mathematical and quantitative technique to make out the decisions being taken. **The goal of OR is to get best decision making by understanding system behavior and improve the system performance.** *The general concept of OR has many applications, for instance, in production management, data analysis, hospital, financial planning, environmental management, inventory control, risk management, military operations, production management, decision &optimization, game theory, telecommunications, engineering system, agriculture, man resource power, sequencing and scheduling, etc. Inventory management (IM) or inventory control is one of the important branches of operations research (OR) mainly used in production management, industry, marketing department, etc. An inventory deals with a decision that minimizes the total average cost or maximizes the total average profit in presents of good customer-seller (retailer) relation. It is very helpful for seller to provide continuous supply of product to the customer. In ordinary inventory model it considers all parameter like shortage cost, holding cost, setup cost as a fixed. But in real life situation it will have some little fluctuations, so consideration of fuzzy parameters is more realistic.*

1.1   Operations Research (OR): Some Basic Concepts

Operations research (OR) is a broad area of applied mathematics which encompasses many diverse area of minimization and optimization. The central work of operations research (OR) is optimization, i.e., "to get best results under the given circumstances". The main origin of operations research (OR) or operational research was world war-2 (1939-1945). At the time of world war-2, limited military resources was decided to use in most effective way. After world war-2, the techniques began to be applied more widely to problems in society, industry, business, etc. This decision was taken-disciplinary terms of scientist to undertake scientific research into strategic and tactical military operations. Operations research (OR) is one of the important discipline that deals with the application of advanced analytical methods to take better decisions. Not only operations research (OR) includes solve the specific problem but also it designing problem solving and implementation systems. Now operations research (OR) is used in every area of business, industry and government throughout the world.

1.1.1   Some historical development of OR

1937 - Modern operations research (OR) originated at Bawdsey research station in UK.

1947 - First mathematical technique in this field (Simplex method) was formed by the American mathematician **George Dantzig.**

1949- Operations research (OR) came into existence at regional research laboratory in Hyderabad.

1950- Operations research achieved recognition as a subject worthy of academic study in universities.

1957- The International federation of operations research (OR) societies was established

1957- Operations research (OR) society of India was formed.

1.1.2   Definition of OR

There are many definitions of OR are available. The following are a few of them.

i)   Operations research (OR) is an analytical procedure of problem-solving and decision-making that is useful in the management of organizations. In OR, problems are broken down into basic components and then solved in defined steps by mathematical analysis.

ii) Operations research (OR) is the application of the theories of probability, linear programming, queuing theory etc., to the probability of war and industry.

iii) Operations research (OR) is the art of winning was without actually fighting (Aurther Clarke).

iv) Operations research (OR) is art of giving bad answers to problems where otherwise worse answers are given (T.L. Saaty.)

v) Operations research (OR) is applied decision theory. It uses any scientific, mathematical or logical means to attempt to cope with the problems that confront the executive when he tries to achieve a thorough going rationality in dealing with the decision problems (D.W. Miller and M.W. Starr).

vi) Operations research (OR) is the application of scientific methods to problems arising from the operations involving integrated system by man, machine and materials (Fabrycky and Torgersen).

1.1.3 Scope of OR

OR has the widely scope in Defense, Industry, Production Department, Marketing Department, Financial Department, L.I.C, Agriculture Planning, etc. It is also closely related to system analysis, management science, control theory, game theory, constraint logic programming, artificial intelligence, multi-criteria decision analysis, and so on.

1.1.4 Some topics of OR

OR as a field, that always tried to maintain its multidisciplinary character and uniqueness. It compromise of various branches. The following are some of them.

i) Assignment problem
ii) Decision analysis
iii) Dynamic programming
iv) **Inventory theory**
v) Queuing theory
vi) Simulation technique
vii) System analysis
viii) System thinking
ix) Game theory
x) Dynamic programming

1.1.5 Solution procedure of OR models

There are three types of methods for solving OR model as follows,

i)   **Analytic procedure** – Differential calculus and finite differences are the analytic procedure for solving OR models. In this method, a general solution has specified by a symbol and the optimal solution can be obtained in a non iterative manner.

ii)   **Numerical procedure** – For this technique, solution procedure start with a trial solution and a set of rules process here for improving it. In this solution method, instead of solving the problem directly, a general algorithm is applied to obtain a specific numerical solution.

iii)   **The Monte-Carlo technique or Simulation** – For this technique, some random numbers are required which may be converted into random variables whose character or behavior is known from past experience. In this case, random samples of specified random variables are drawn to know what is happening to the system for a selected period of time under different conditions.

1.2   Inventory Management

An inventory management (IM) is the set of policies and controls that monitor the levels of inventory and determine what levels should be maintained, when the stock should be replenished, and how large the orders should be stock of an item. For this purpose the task is to construct a mathematical model of the real life inventory system, such a mathematical model is formed based on various assumption and approximation. Successful key of a business is to fulfillment customer demand within shortest possible time with the best quality, and competitive price. Using inventory management (IM) this is possible to achieve. Inventory model (or more formally the mathematical theory of inventory and production) is the sub-specially with in operations research (OR) or operations management (OM) that is concerned with the design of production system to minimize total average cost. It studies the decision faced by firms and the military in connection with manufacturing, warehousing, supply chains, spare part allocation and so on and provides the mathematical foundation for logistics. The inventory control problem is the problem faced by a firm that must decide how much to order in each time period to meet demand for its products. The problem can be modifying using mathematical techniques of optimal control, dynamic programming and network optimization. The study of such models is part of inventory theory or inventory model. An inventory for a manufacturing facility consists of three major types. Raw materials are the basic inputs to the manufacturing process. Work in process consists of the partially finished goods. Finished of goods are the outputs of the manufacturing process.

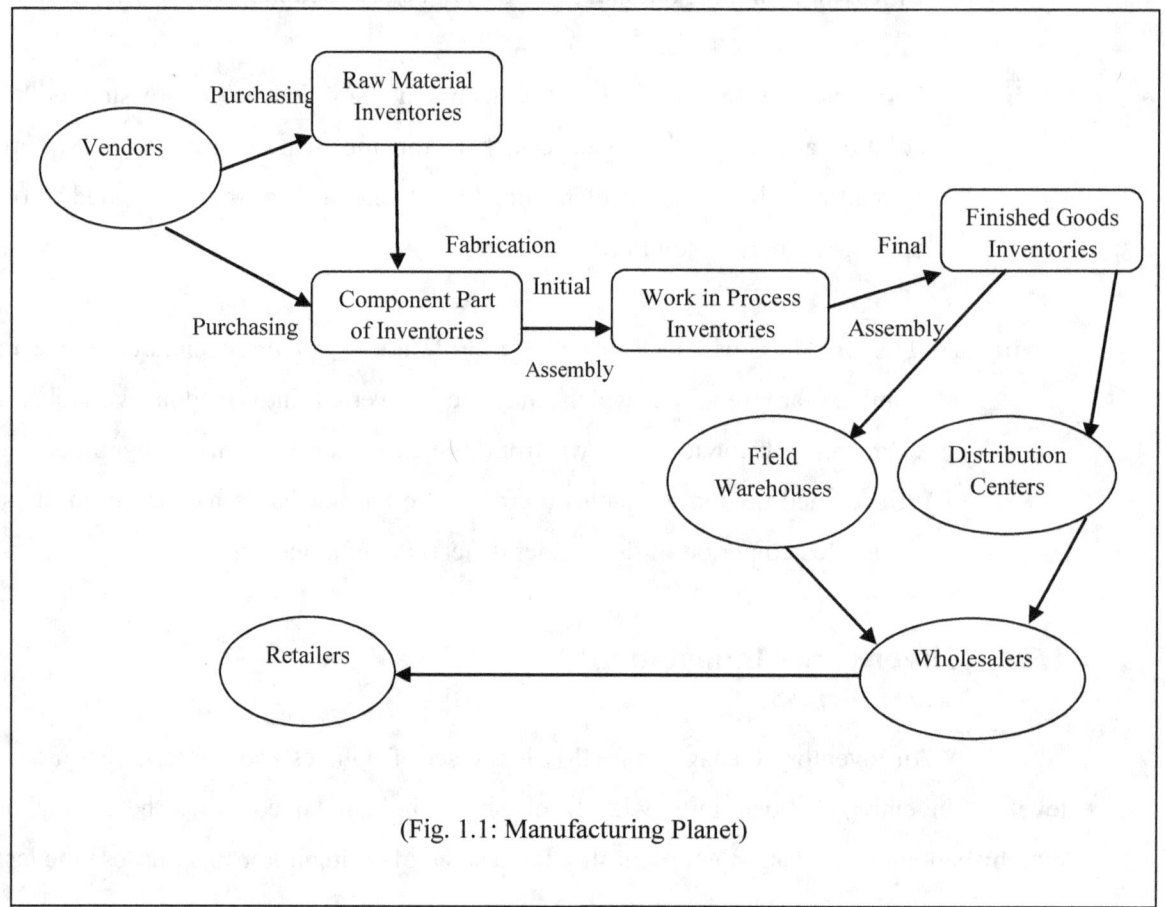

(Fig. 1.1: Manufacturing Planet)

In the past, researchers assumed the parameters involved in EOQ model such as the demand, holding cost, se-up cost, deteriorating cost, etc. as a crisp values or random variables. But in reality, the demand and cost of the items change from time to time. Again, calculating these variables by the probability distribution is very hard due to lack of historical data. The cost parameters generally estimated based on previous experiment and managerial judgment. Therefore, in real life situation fuzzy set theory is more realistic than crisp set theory or traditional probability theory.

Now, in the context of making decisions, Bellman and Zadeh quoted: *"Much of the decision making in the real-world takes place in an environment in which the goals, the constraints and the consequence of possible actions are not known precisely"*. Fuzziness is the ambiguity that can be found in the definition of a concept or the meaning of a word.

Actually, everyone is involved with fuzziness and it is a kind of uncertainty that anyone can understand.

The fuzzy set theory was introduced by L.A. Zadeh in 1965.Bellman & Zadeh (1970) applied fuzzy set theory to the decision-making problem. Later, Tanaka et al. (1974) presented the objectives as fuzzy goals over the α − cuts of a fuzzy constraints set and in 1976 Zimmermann showed that the classical algorithms could be used to solve a fuzzy linear programming problem (FLPP) and fuzzy additive programming technique. The literature on fuzziness to real life problems is well covered in the book prepared by Sangalli in 1998. Randomness and fuzziness differ in nature and fuzziness expresses much more everyday uncertainty than probability.

1.2.1   Types of inventory

An inventory exists only because there is a difference in timing between supply and demand. For example, if received a product that was immediately (instantaneous) demanded, there would be no need to store it, therefore no inventory occurs here. As such, when the rate of supply is exceeds the rate of demand, inventory increases; when the rate of demand is exceeds the rate of supply, inventory decreases. However, there are various reasons for an imbalance between the rates of supply and the rate of demand. Now the different types of inventory as follows

- **Raw materials:** This type of inventory includes any goods or products used in the manufacturing process, such as components used to assemble a finished product. Partially finished goods or materials are also raw materials. For example, for an orange juice company, oranges, sugar and preservatives are raw materials.

- **Lot size inventory:** This inventory exists when there is more production and less demand. Amount of inventory depends upon investment, storage space, shipment quantity, resource etc.

- **Work-in-process:** Work-in-process inventory items are those goods or materials that are waiting to be made into something else. These may include partially assembled materials that are waiting to be completed.

- **Finished goods:** Finished goods are any items or products that are ready to sell directly to customers, including to wholesalers and retailers.

- **Buffer inventory:** Buffer inventory also referred to as a safety inventory. Its purpose is to compensate for unexpected fluctuations in supply of demand.

- **Cycle inventory:** Cycle inventory occurs because one or more stages in the process cannot supply all the items it produces simultaneously. This type of inventory only results from the need to produce products in batches and the amount of it depends on volume decisions.

- **Pipeline inventory:** It exists because products cannot be transported instantaneously between the point of supply and the point of demand.

- **Anticipation inventory:** It is the inventory that is accumulated to cope with expected future demand or interruptions in supply.

- **Seasonal inventory:** This type of inventory is created to meet demand which is caused due to seasonal variation in demand.

- **Decoupling inventory:** When various manufacturing process operates successively then failure of any one can interrupt whole the production process. To overcome this stocking point of the inventory takes place between adjacent stages.

- **Fluctuation inventory:** It is the inventory that acts equilibrium between sales and production. The reserve stock that is kept to maintain fluctuations in the demand and lead – time, affecting the production of the items is called fluctuation inventory.

1.2.2　Inventory decision

Generally an inventory problem starts with three questions:

- What item should be kept in stock?
- When to order?
- How much order?

And inventory decision taken as following chart

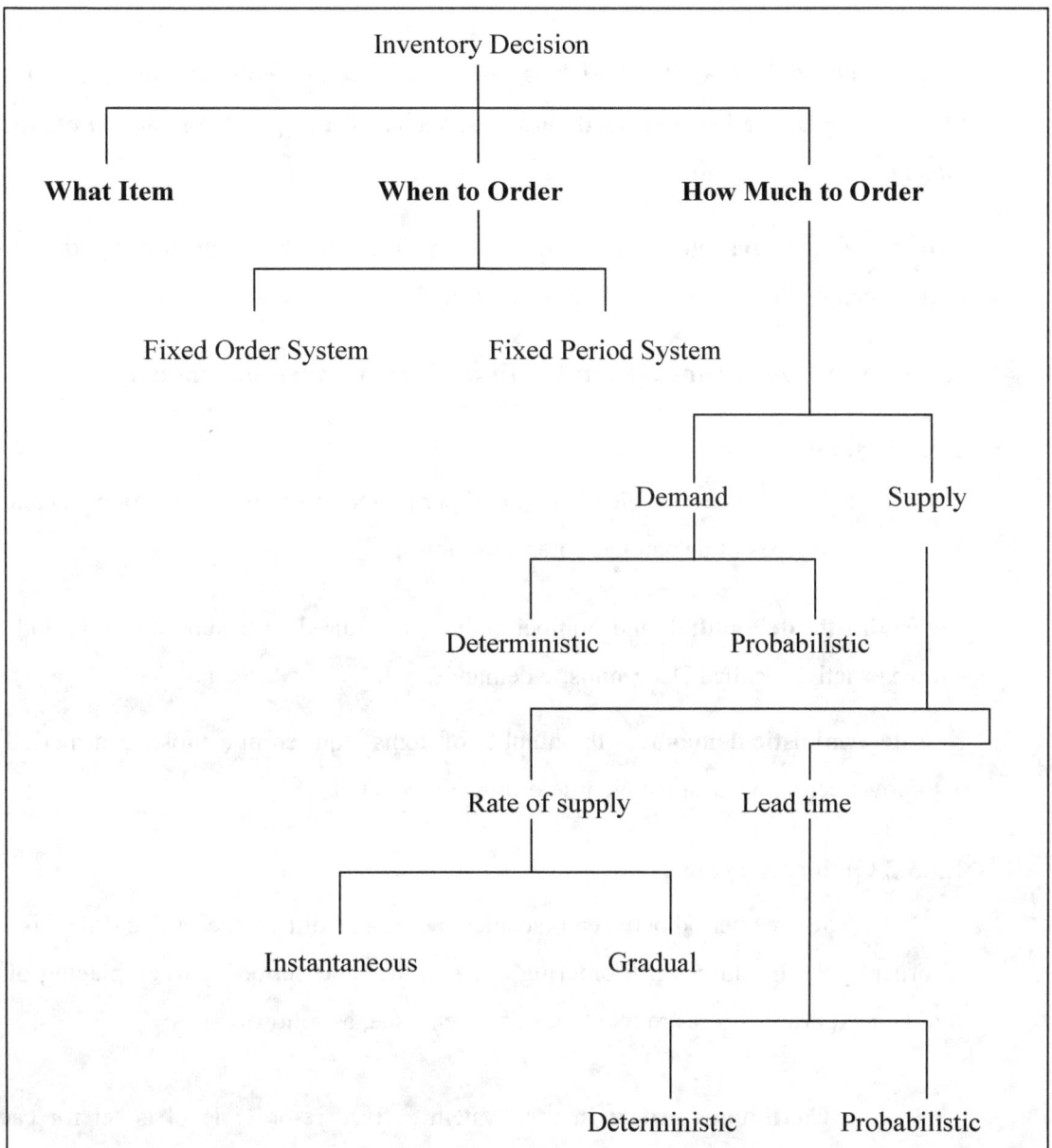

1.2.3   Definition of inventory

The stock of goods or commodities which is held to fulfill the future dream is called an inventory. In other word an inventory is a mathematical equation or formula that helps a company or firm in determining the economic order quantity, and the frequency of ordering, to keep goods or services flowing to the customer without interruption or delay. In general sense, an inventory is the raw materials, work-in-process products and finished goods that are considered to be the portion of a business's assets that are ready or will be ready for sale.

1.2.4   Variables in an inventory model

Variables in an inventory model as follows:

Controlled variables: The variables which can be controlled by the stock matter through-out the inventory are called controlled variables, such as the stage of completion of stocked items, money, etc.

Uncontrolled variables: The variables which cannot be controlled by the stock matter through-out the inventory are called uncontrolled variables, such as demand.

1.2.5   Some basic concepts and definition of an inventory model

1.2.5.1 Demand

Number of items which is required per period in an inventory model is called demand. There are two types of demand, defined as follows;

Deterministic demand: If the number of items required in a subsequent period of time is known exactly, is called Deterministic demand.

Non-deterministic demand: If the number of items required in a subsequent period of time is not known certainty, is called Non-deterministic demand.

1.2.5.2 Ordering cycle

The time period between placements of goods of two successive orders is called to as an order cycle. In other word, ordering cycle is the time period between placing of one order and the next order. There are two types of order cycle, as follows

i)   **Continuous order:** In this system a fixed re-order level is set for each item of goods and inventory level is supervised continuously.

ii)   **Periodic order:** In this system, the inventory time divided by equal intervals and orders of variable size are placed at regular interval of time.

1.2.5.3 Lead time

Latency between the initiation and execution of process is called lead time. In other hand lead time is the length of delay when an order of goods is placed and the date goods are available for use. Lead time plays an important role between seller-customer relations. In inventory or supply chain management from customer order received to the moment of the ordered delivered divided into the five lead times as follows:

i)   **Order lead time** - Time from customer order received to customer order delivered.

ii)   **Order handling time** - Time from customer order received to sales order crated.

iii)   **Manufacturing lead time** – Time from order created to production finished.

iv)   **Production lead time** – Time from start of physical production of first sub-module.

v)   **Delivery lead time** – Time from production finished to customer order delivered.

Here we have given a rough figure of Lead time in Fig. 1.2.

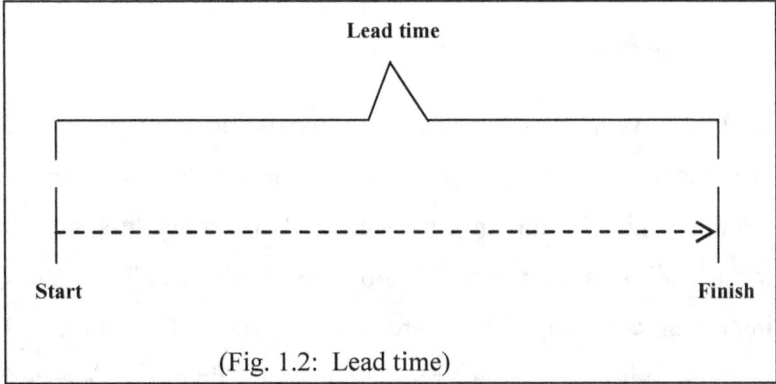

(Fig. 1.2: Lead time)

1.2.5.4 Time horizon

This is defined as a time period over which a particular inventory model is maintained. This can be finite or infinite depending on the demand.

1.2.5.5 Reorder level

This is initial level at which ordering activities must be initiated as the on hand stock just cover the demand until ordered units in the inventory.

1.2.5.6 Buffer stock

Buffer stock is one of important part of an inventory management (IM). It is defined as a supply of inputs held as a reserve in case there is future demand and supply fluctuations. For uncertain of future buffer stock is excess inventory or safety stock. Buffer stock found at all stages of the inventory management and it reduces stock-out situation. Thus buffer stock provides a better customer service.

1.2.5.7 Maximum stock

For limited capital and limited capacity of ware-house the allowable stock of goods, also be limited in an inventory. The maximum allowable stock in an inventory is called maximum stock.

1.2.5.8 Number of items

Normally, an inventory model involves more than one commodity. The number of items in an inventory affects for limited floor capacity or limited total capital.

1.2.5.9 Deterioration

In an inventory model deterioration plays an important role. Deterioration is defined as decay or damage in the quality of the inventory. Foods, Drugs, pharmaceuticals etc. are deteriorating items. During inventory there have some losses of these deteriorating items; consequently this loss must be taken into account when analyzing the system. Shortages are also very important condition. There are several types of customer. At shortage period some customers are waiting for actual product and others do not it. Deterioration may happen due the following reasons.

- The items or products may have a fixed life, e.g., bulbs, photo films, medicines.
- Poor handling the store, e.g., fruits and vegetable, crockery items.
- Unsatisfactory, poor or inadequate storage conditions, e.g., dairy products.

1.2.5.10 Stock-Out/Out-Of-Stock (OOS)

A situation in which the demand or requirement for an item cannot be fulfillment properly from the current inventory is called stock-out or out-of stock (OOS). Stock-out condition arises in an inventory model due to unexpected demand, ineffective inventory management, production delays or replenishment disruptions. For stock-out companies may loss of future business due to customer dissatisfaction.

1.2.5.11 Inflation

Inflation is the rate at which the general level of prices for products or goods and services is rising and, therefore, the purchasing power of currency is falling over a period of time. But it should be know that the inflation refers to the general trend of prices, not changes any specific price. Inflation plays very important roles in presented inventory system. In past many researchers was neglecting it, but consideration of inflation inventory model is being more realistic.

1.2.5.12 Shortage

Shortage plays very important roles in an inventory model. Shortages comes at the situation where the quantity available or supplied in a market falls short of the quantity demanded or required at a given time or price. Shortages of stocks or stock-out time of an inventory model may result in the cancellation of orders and heavy losses in sale which in turn may result in loss in goodwill, profit even the business itself of a company. When a stock out or shortage occurs in an inventory, a company or seller faces two possibilities:

•It can meet the shortage with some type of rush, special handling or priority shipment.

•It cannot meet the shortages at all.

The cost associated with a stock out or shortage depends on how the company or seller handles the problem.

1.2.5.13 Stock replenishment

An inventory problem may operate with lead time but the actual replacement of stock may occur instantaneously or uniformly. Instantaneous replenishment occurs due the stock is purchased from outside sources whereas the uniform replenishment may occur when the product is manufactured by the company.

1.2.5.14 Standardization

In an inventory control, standardization is the determination of fixed sizes, shapes, quality and dimensions of material. As it is desirable to encourage the use of standards, a great deal of reduction is possible in inventory space, obsolescence, handling costs and inventory as well as improvement of the quality.

1.2.5.15 EOQ model

Economic order quantity (EOQ) model is one of the oldest models that minimize the total holding costs and ordering cost. This is also known as Wilson EOQ model or Andler formula. Ford W. Harris was developed this model in 1913 bur R.S. Wilson applied it extensively and K. Andle presents depth analysis on it. EOQ model is applied when demand is constant over the given time period and each new order is delivered when inventory reaches zero. In an EOQ model there is a fixed cost for each order placed also there is a cost for each unit held in storage, known as holing cost or carrying cost. In an inventory we want to determine the optimal number of units to order so that we minimize the total average cost.

In an EOQ model variables are

- Unit production cost
- Setup cost
- Annual holding cost per unit
- Order quantity
- Optimal order quantity
- Annual demand quantity

In a single item EOQ formula, our objects to minimize the following cost function:

Total Cost = production cost + ordering cost + holding cost.

Where

Production cost = Purchase unit price × annual demand quantity = $P \times D$

Ordering cost = Cost of placing of orders or items = $K \times \frac{D}{Q}$.

Holding cost = Holding cost per unity × the average quantity in a stock = $h \times \frac{Q}{2}$.

Here,

K = ordering cost,

D = demand rate,

h = holding cost per unit,

P = unit production cost,

Q = order quantity.

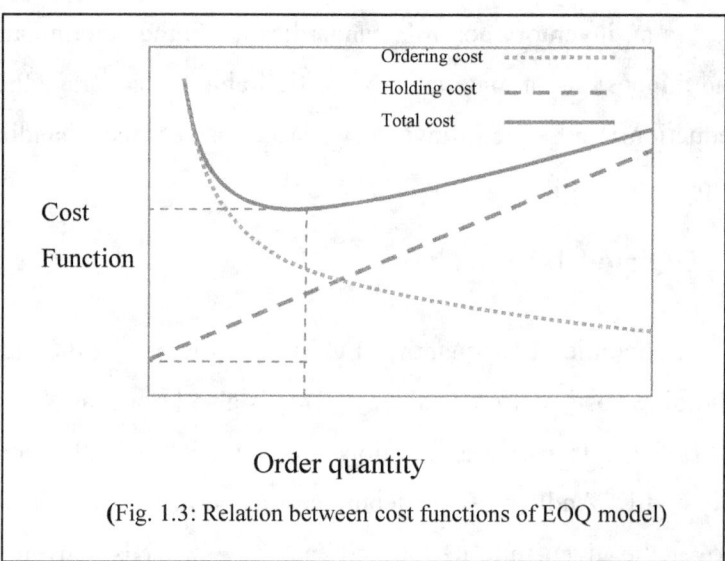

(Fig. 1.3: Relation between cost functions of EOQ model)

1.2.5.16 EPQ model

The economic order quantity (EOQ) model determines the quantity a company should order to minimize the total average cost by balancing the holding cost and fixed ordering cost. In 1918 E.W. Taft was developed the economic production quantity (EPQ) model. This model is extension of EOQ model. The difference of EOQ and EPQ model is that the EOQ model assumes the order quantity arrives completely and immediately after it ordering, but EPQ model assumes the company or retailer will produce its own quantity. An EPQ model applies where the demand for a production is constant over a time period and a new order is placed when inventory reaches zero. In an inventory there is a fixed cost for each order placed. There is also a storage cost for each unit is called holding cost or carrying cost. In an inventory we want to determine the optimal number of units to order so that we minimize the total average cost associated with the fixed, delivery, purchase and storage of the product.

In an EPQ model variables are

- Unit production cost
- Setup cost
- Annual holding cost per unit
- Order quantity
- Optimal order quantity
- Annual demand quantity
- Demand rate
- Production rate

In a single item EPQ formula, our objects to minimize the following cost function:

Total Cost = production cost + ordering cost + holding cost.

Where

Production cost = Purchase unit price × annual demand quantity = $P \times D$.

Ordering cost per year = Cost of placing of orders or items = $K \times \frac{D}{Q}$.

Holding cost per year = Holding cost per unity × the average quantity in a stock = $h(1-x)\frac{Q}{2}$.

K = ordering cost,

D = demand rate,

h = holding cost,

T = cycle length,

P = production rate,

x = D/P,

Q = order quantity.

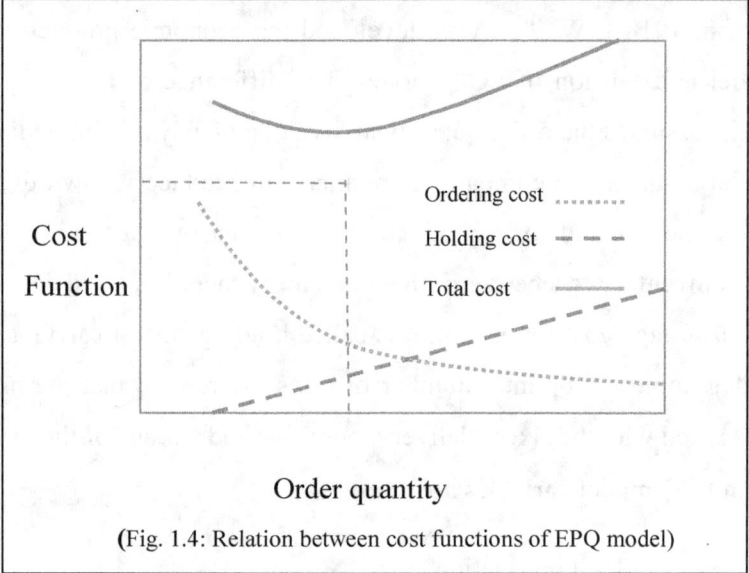

(Fig. 1.4: Relation between cost functions of EPQ model)

1.2.6 Cost variables in inventory model

In an inventory generally three type of cost functions, as follows:

Holding cost: The costs which are controlled in holding the inventory, are called holding cost. Various type of Holding cost as follows,

       a. Capital investment.

       b. Interest of capital.

       c. Labor charges for maintaining stock.

       d. Insurance and taxes.

       e. Purchase price or production costs etc.

Shortages cost: The second kind of cost that of incurring shortages, is the cost of lost sales, of loss good will, of overtime payments, of special administrative efforts (telephone calls, memos, and letters), etc. The unsatisfied demand or shortages penalty cost is incurred when the stock proves to be inadequate to meet the demand of customers.

Set-up cost: The third kind of cost that of replenishing inventories, is the cost of machine setups for production runs, of preparing orders, of handling shipment, etc. The cost associated with obtaining goods through placing of an order or purchasing or manufacturing or setting up machinery before starting production.

1.2.7 Total inventory cost

Total inventory cost (TC) is the sum of holding cost (c_1), shortages cost (c_2), and set-up cost (c_3). Thus, the total cost (TC), is TC $= c_1 + c_2 + c_3$. If, T is the cycle of time or time period, then total average cost (TAC), is given by TAC $= \frac{c_1 + c_2 + c_3}{T}$.

1.2.8 Types of inventory systems

An inventory model or system has been defined as a system in which two or all kinds of cost are subject to control. Now distinguish among several types of inventory systems as follows:

In a type (1, 2) inventory system only the holding cost and the shortages cost are subject to control. In a type (1, 3) inventory system only the holding cost and the replenishing cost are subject to control. Similarly in a type (2, 3) inventory system only the shortages cost and the replenishing cost are subject to control. In such system the holding cost is not subject to control. A type (1, 2, 3) inventory system is a system in which all three costs are subject to control By using this classification, inventory system can be grouped into four classes: type (1, 2), type (1, 3), type (2, 3), and type (1, 2, 3). We shall see that there are practically no difference between type (1, 2) and type (2, 3) inventory systems. We shall also that in some circumstance type (1, 2) and type (1, 3) systems are special case of type (1, 2, 3) systems. Since this is not true in general. We can-not content ourselves with the study of type (1, 2, 3) systems alone but will have to study the other types of systems as well.

1.2.9 Importance of inventory management

Many companies or retailers underestimates the importance of inventory management which proves to hazardous but inventory model or inventory management plays very importance role in making business and industry. The good inventory management knowledge provides better business decision. Important of inventory management as follows:

i)   It keeps better seller-customer relations.

ii)   It gives better utilization of man machinery.

iii)   It allows possible increase in output.

iv)   It maintains smooth and efficient production flow.

v)   It gives to take advantage of quantity discount.

vi)   It utilizes to advantage price fluctuations.

vii)   It helps in smooth and effective running of business.

viii)   It improves the man power, equipment and facility utilization because of better planning.

ix)   It helps to minimizing the loss due to deterioration, obsolescence damage etc.

x)   It takes advantages of price discounts by bulk purchasing.

1.3   The Theory of Fuzzy Sets: Some Basic Concepts

Fuzzy sets were introduced by L.A. Zadeh in 1965.The fuzzy set theory is developed to improve the oversimplified model, thereby developing a more robust and flexible model in order to solve real world complex systems involving human aspects. Furthermore, it helps the decision maker not only to consider the existing alternatives under given constraints (optimizing a given system) but also to develop new alternatives (design a system). The theory of fuzzy sets has appeared as the most promising foot for dealing with human centered decision making problems under Fuzziness. The Fuzzy set theory expresses Fuzziness by means of the concept of the sets in the traditional sense. It provides a strict mathematical framework in which vague conceptual phenomenon can be precisely and rigorously studied. It can also be considered as a modeling language for describing Fuzzy criteria, phenomenon and Fuzzy relations existing in real-life situations. The Fuzzy set theory, however, abandons the excluded-middle law and the law of contradiction of non-fuzzy sets, and thus abandons the standards probability calculus altogether, since it destroys the De Morgan relations.

Now some formal definitions in the framework of Fuzzy set theory and discussed as follows.

1.3.1 Definition of fuzzy sets

Fuzzy set $\tilde{A}$ in X is defined by a set of orders pairs:

$$\tilde{A} = \{(x, \mu_{\tilde{A}}(x))| x\epsilon X\}.$$

Where $\mu_{\tilde{A}}(x)$ represents the grade membership of x in $\tilde{A}$.

The function $\mu_{\tilde{A}}(x)$ defines the degree to which the element x of the set X is included in $\tilde{A}$.

The degree inclusion is also called degree of truth.

The function $\mu_{\tilde{A}}(x)$ may also be express as:

$$\mu_{\tilde{A}}(x):X \longrightarrow [0, 1].$$

Thus the value of $\mu_{\tilde{A}}(x)$ nearer to unity implies the higher grade of membership of x in $\tilde{A}$ and vice-versa. When $\mu_{\tilde{A}}(x)$ contains only two points 0 and 1, then $\mu_{\tilde{A}}(x)$ is identical to the characteristic function

$$C_A: X \longrightarrow \{0, 1\}.$$

Where A is an ordinary crisp set and hence $\tilde{A}$ is not fuzzy set.

An ordinary set A is conventionally expressed as:

$$A = \{x \epsilon X \mid C_A(x) = 1\},$$

Through its characteristic function

$$C_A(x) = \begin{cases} 1 & if \quad x \in A, \\ 0 & if \quad x \notin A. \end{cases}$$

Thus membership function is an obvious extension of the idea of a characteristic function of an ordinary set, because it takes values not only 0 and 1 but also the values in between them.

1.3.1 Theoretic operations on fuzzy sets

Several operators for fuzzy sets have been developed. Some of the basic operators originally suggested by L.A. Zadeh (1965), are presented as follows:

• **Union**

Union of two Fuzzy sets $\tilde{A}$ and $\tilde{B}$ on X, denoted by $\tilde{A}U\tilde{B}$, is defined by

$$\mu_{\tilde{A}U\tilde{B}} = \max\{\mu_{\tilde{A}}(x), \mu_{\tilde{B}}(x)\}, \text{ for all } x\epsilon X.$$

• **Intersection**

The intersection of two fuzzy sets $\tilde{A}$ and $\tilde{B}$ on X, denoted by $\tilde{A}\wedge\tilde{B}$, is defined by

$$\mu_{\tilde{A}\cap\tilde{B}} = \min\{\mu_{\tilde{A}}(x), \mu_{\tilde{B}}(x)\}, \text{ for all } x\epsilon X.$$

• **Complement**

The compliment of fuzzy set $\tilde{A}$, denote by $\tilde{A}^C$ is defined by

$$\tilde{A}^C = 1 - \mu_{\tilde{A}}(x), \text{ for all } x \in X.$$

• Equality

Two sets $\tilde{A}$ and $\tilde{B}$ on X are said to be equal, denoted by $\tilde{A} = \tilde{B}$, if and only if their membership values at each   point $x \in X$ are equal. i.e.,

$$\tilde{A} = \tilde{B} \leftrightarrow \mu_{\tilde{A}}(x) = \mu_{\tilde{B}}(x), \text{ for all } x \in X.$$

• Inclusion

A Fuzzy set $\tilde{A}$ is contained in the Fuzzy set $\tilde{B}$ on X, denoted by $\tilde{A} \subseteq \tilde{B}$ if and only if the membership function values of all x belongs to $\tilde{A}$ is less than or equal to that of all x belongs to the $\tilde{B}$, i.e,

$$\tilde{A} \subseteq \tilde{B} \leftrightarrow \mu_{\tilde{A}}(x) \leq \mu_{\tilde{B}}(x), \text{ for all } x \in X.$$

1.4   Fuzzy Number

A fuzzy number is a quantity whose value is imprecise, rather than exact as is the case with single valued number. A fuzzy number dose not refer to one single value it is a connected set of possible values, where each possible values has a weight between 0 and 1. This weight is called membership grade or membership function. Thus membership function of a fuzzy set, which have the form

$$\mu_{\tilde{A}}(x): R \rightarrow [0, 1].$$

Where $\mu_{\tilde{A}}(x)$ is a membership function of the fuzzy set $\tilde{A}$.

To qualify as a fuzzy number, a fuzzy set $\tilde{A}$ on R must possess at the following properties:

i) $\tilde{A}$ must be normal fuzzy set;

ii) A_α must be closed interval for every $\alpha \in (0, 1]$;

iii) The support of $\tilde{A}$, $A_0{}^+$ must be bounded;

A real number $\tilde{A}$ described as fuzzy subset on the real line $\mathcal{R}$ whose membership function $\mu_{\tilde{A}}(x)$ has the following characteristics with $-\propto < a_1 \leq a_2 \leq a_3 < \propto$

$$\mu_{\tilde{A}}(x) = \begin{cases} \mu_{\tilde{A}}^L(x) & if\, a_1 \leq x \leq a_2, \\ \mu_{\tilde{A}}^R(x) & if\, a_2 \leq x \leq a_3, \\ 0 & otherwise. \end{cases}$$

Where $\mu_{\tilde{A}}^L(x): [a_1, a_2] \rightarrow [0,1]$ is continuous and strictly increasing and $\mu_{\tilde{A}}^R(x): [a_2, a_3] \rightarrow [0,1]$ is continuous and strictly decreasing.

1.5   Generalized Fuzzy Number (GFN)

Generalized Fuzzy Number (GFN): Chen represents a generalized trapezoidal fuzzy number (GTrFN) $\tilde{A}$ as $\tilde{A} = (a, b, c, d; w)$, where $0 < w \leq 1$, and a, b, c and d are real numbers. The generalized fuzzy number (GFN) $\tilde{A}$ is a fuzzy subset of real line R, whose membership function $\mu_{\tilde{A}}(x)$ satisfy the following properties

a) $\mu_{\tilde{A}}(x)$ is continuous mapping from R to the closed interval [0, 1];

b) $\mu_{\tilde{A}}(x) = 0$ for all x$\in (-\infty, a)$;

c) $\mu_{\tilde{A}}(x)$ is strictly increasing with constant rate on [a, b];

d) $\mu_{\tilde{A}}(x) = w$ for all x$\in$[b, c];

e) $\mu_{\tilde{A}}(x)$ is strictly decreasing with constant rate on [c, d];

f) $\mu_{\tilde{A}}(x) = 0$ for all x$\in (d, \infty)$;

Note: $\tilde{A}$ is a normalized fuzzy number when w = 1, and it is non-normalized for w $\neq$ 1.

1.6   Generalized Trapezoidal Fuzzy Number (GTrFN)

Generalized Trapezoidal Fuzzy Number (GTrFN): A generalized fuzzy number (GTrFN) $\tilde{A} = (a, b, c, d; w)$ is a fuzzy set of the real line R whose membership function $\mu_{\tilde{A}}(x)$: R $\rightarrow$ [0, w] is defined as

$$\mu_{\tilde{A}}^W(x) = \begin{cases} \mu_{L\tilde{A}}^W(x) = w\left[\dfrac{x-a}{b-a}\right], & a \leq x \leq b; \\ 1 & , & b \leq x \leq c; \\ \mu_{R\tilde{A}}^W(x) = w\left[\dfrac{x-d}{c-d}\right], & c \leq x \leq d; \\ 0 & , & otherwise; \end{cases}$$

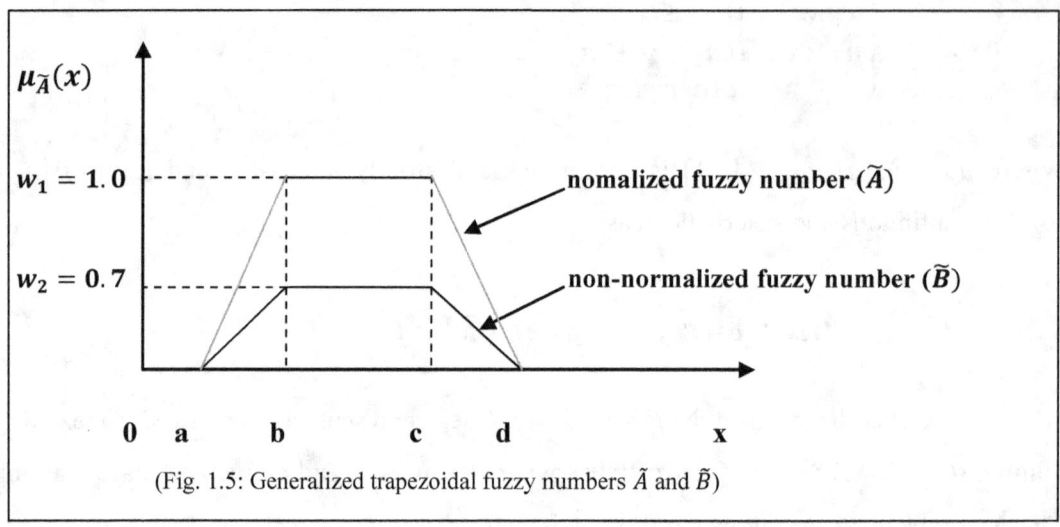

(Fig. 1.5: Generalized trapezoidal fuzzy numbers $\tilde{A}$ and $\tilde{B}$)

Where $\mu_{L\tilde{A}}{}^{W}(x)$ and $\mu_{R\tilde{A}}{}^{W}(x)$ are the left and right membership function of $\tilde{A}$ respectively. The inverse functions $\mu^{-1}{}_{L\tilde{A}}{}^{W}(x): [0, w] \to [a, b]$ and $\mu^{-1}{}_{R\tilde{A}}{}^{W}(x): [0, w] \to [c, d]$ are defined as

$$\mu^{-1}{}_{L\tilde{A}}{}^{W}(y) = a + \frac{y(b-a)}{w};$$
$$\mu^{-1}{}_{R\tilde{A}}{}^{W}(y) = d + \frac{y(c-d)}{w}; \qquad y \in [0, w].$$

For a non-normal fuzzy number $\tilde{A}$, the corresponding membership function $\mu_{\tilde{A}}(x)$ can be normalized by dividing the maximal value of $\mu_{\tilde{A}}(x)$ i.e., w.

1.7   α-level Set

The α- level of a fuzzy number is defined as a crisp set where A(α) = [x: $\mu_{\tilde{A}}$(x) ≥ α, x∈X] where α∈ [0,1]. A(α) is a non-empty bounded closed interval contained in X and it can be denoted by A_{α} = [A_L(α), A_R(α)]. A_L(α) and A_R(α) are the lower and upper bounds of the closed interval, respectively.

1.8   Interval Number

An interval number A is defined by an ordered pair of real numbers as follows A = $[a_L, a_R]$ = {$x: a_L \leq x \leq a_R$}, $x \in \mathcal{R}$} where a_L and a_R are the left and right bounds of interval A, respectively. The interval A, is also defined by center (a_c) and half-width (a_w) as follows; A = (a_c, a_W) = {$x: a_c - a_w \leq x \leq a_c + a_w$, $x \in \mathcal{R}$} where $a_c = \frac{a_R + a_L}{2}$ is the centre and $a_w = \frac{a_R - a_L}{2}$ is the half-width of A.

1.9 Nearest Interval Approximation (NIA)

Here we want to approximate a fuzzy number by a crisp model. Suppose $\tilde{A}$ and $\tilde{B}$ are two fuzzy numbers with α-cuts are $[A_L(\alpha), A_R(\alpha)]$ and $[B_L(\alpha), B_R(\alpha)]$, respectively. Then the distance between $\tilde{A}$ and $\tilde{B}$ is $d(\tilde{A}, \tilde{B}) = \sqrt{\int_0^1 (A_L(\alpha) - B_L(\alpha))^2 + \int_0^1 (A_R(\alpha) - B_R(\alpha))^2 d\alpha}$.

Given a fuzzy number $\hat{A}$, we have to find a closed interval $C_D(\tilde{A})$, which is closest to $\hat{A}$ with respect to some metric. We can do it since each interval is also a fuzzy number with constant α-cut for all $\alpha \in [0, 1]$. Hence $(C_D(\tilde{A}))\alpha = [C_L, C_R]$.

Now we have to minimize $d(\tilde{A}, C_D(\tilde{A})) = \sqrt{\int_0^1 (A_L(\alpha) - C_L)^2 + \int_0^1 (A_R(\alpha) - C_R)^2 d\alpha}$ with respect to C_L and C_R. In order to minimize $d(\tilde{A}, C_D(\tilde{A}))$, it is sufficient to minimize the function $D(C_L, C_R) = (d^2(\tilde{A}, C_D(\tilde{A})))$.

The first partial derivatives are $\frac{\partial}{\partial C_L} D(C_L, C_R) = -2\int_0^1 A_L(\alpha)d\alpha + 2C_L$ and $\frac{\partial}{\partial C_R} D(C_L, C_R) = -2\int_0^1 A_R(\alpha)d\alpha + 2C_R$. Solving $\frac{\partial}{\partial C_L} D(C_L, C_R) = 0$ and $\frac{\partial}{\partial C_R} D(C_L, C_R) = 0$, we get $C_L = \int_0^1 A_L(\alpha)d\alpha$ and $C_R = \int_0^1 A_R(\alpha)d\alpha$. Again since $\frac{\partial^2}{\partial C_L^2}(D(C_L^*, C_R^*)) = 2 > 0$, $\frac{\partial^2}{\partial C_R^2}(D(C_L^*, C_R^*)) = 2 > 0$ and $H(C_L^*, C_R^*) = \frac{\partial^2}{\partial C_L^2}(D(C_L^*, C_R^*)) \cdot \frac{\partial^2}{\partial C_R^2}(D(C_L^*, C_R^*)) - \left(\frac{\partial^2}{\partial C_L^* C_R^*}(D(C_L^*, C_R^*))\right)^2 = 4 > 0$. So $D(C_L^*, C_R^*)$ i.e., $d(\tilde{A}, C_D(\tilde{A}))$ is global minimum. Therefore, the interval $C_d(\tilde{A}) = \left(\int_0^1 A_L(\alpha)d\alpha, \int_0^1 A_R(\alpha)d\alpha\right)$ is the nearest interval approximation (NIA) of fuzzy number $\tilde{A}$ with respect to the metric d.

Let $\tilde{A} = (a_1, a_2, a_3)$ be a triangular fuzzy number. The α-cut interval of $\tilde{A}$ is defined as $A_\alpha = [A_L(\alpha), A_R(\alpha)]$ where $A_L(\alpha) = a_1 + \alpha(a_2 - a_1)$ and $A_R(\alpha) = a_3 - \alpha(a_3 - a_2)$. By nearest interval approximation method, the lower limit of the interval is $C_L = \int_0^1 A_L(\alpha)d\alpha = \int_0^1 [a_1 + \alpha(a_2 - a1)]d\alpha = a1 + a2\,2$, and the upper limit of the interval is $C_R = \int_0^1 A_R\alpha d\alpha = \int_0^1 [a3 - \alpha(a3 - a2)]d\alpha = \frac{a_3 + a_2}{2}$. Therefore, the interval number corresponding to the fuzzy number $\tilde{A}$ is $[\frac{a_1 + a_2}{2}, \frac{a_3 + a_2}{2}] = [m, n]$. In the centre and half–width form the interval number of $\tilde{A}$ is defined as $\langle \frac{1}{4}(a_1 + 2a_2 + a_3), \frac{1}{4}(a_3 - a_1)\rangle$.

1.10 Parametric Interval - Valued Function

Let [m, n] be an interval, where m > 0, n > 0. From analytical geometry point of view, any real number can be represented on a line. Similarly, we can express an interval by a function. The parametric interval-valued function for the interval [m, n] can be taken as $g(s) = m^{1-s} n^s$ for $s \in [0, 1]$, which is a strictly monotone, continuous function and its inverse exits. Let ψ be the inverse of g(s), then

$$s = \frac{log \psi - log m}{log n - log m}.$$

1.11   Literature Review

F.W. Harris in 1913 was earliest known analysis of an inventory system. Harris was the first person who published the classical lot size formula

$$q_0 = \sqrt{\frac{2rc_3}{c_1}}. \qquad (1.1)$$

Where q_0 is the optimal lot size quantity, r is the rate of demand per unit time, c_1 is the carrying cost per unit inventory per unit of time and c_3 is the cost of replenishing inventory. Ford W. Harris (1913) was developed this model but R.S. Wilson applied it extensively and K. Andle presents depth analysis on it. In **1918** E.W. Taft was developed the EPQ model. Benjamein Cooper in 1926, who first analyzed an inventory system in which the rate of production was considered. In 1928 Thornton C. Fry presented an inventory system in which requirements were not known precisely. In 1931 Fairfrekd E. Raymond, who first attempt to deal with large variety of inventory systems and to present the beginning of the theory of inventory systems in his research paper **Quantity and Economy Manufacture.** After the World War II (1939-1945) the study of inventory management (IM) has increased rapidly. In 1953 an excellent review of the inventory model (IM) that was studied until1951 is given by Whitin's bibliography in *"The Theory of Inventory Management (IM)"*. The paper "Optimal Inventory Policy" by Arrow, Harris and Marschak in 1951 had given the idea of what may be called the modern analysis of inventory systems. In 1952 Dvoretsky, Kiefer and Wolfowitz in their two papers on "The Inventory Problem" analyze the general systems of inventory policy proposed by Arrow, Harris and Marschak. In these papers very powerful mathematical and statistical tools were introduced, and systems presented were of a very general in nature. After 1952 more and more research work has been devoted to inventory systems. Many research paper or article on the subject now appear regularly in Operations Research (OR), Management Science, Logistics Quarterly,

Journal of Industrial Engineers, Naval Research, European Journal of Operational Research, Annals of Operations Research and many other journals.

The study of inventory model where demand rates varies with time is in last century. Friedman (1978) presented continuous time inventory model with time varying demand. Park (1982) presented an inventory model with partially backorders. M. Roychowdhury and K.S Chaudhuri (1983) studied an order level inventory for deteriorating items with finite rate of replenishment. Ritchie (1984) studied in inventory model with linear increasing demand. Goswami, Chaudhuri (1991) discussed an inventory model with shortage. Datta and pal (1991) investigated an inventory system with power demand pattern and deterioration. A.M. Hariga (1996) presented optimal EOQ models for deteriorating items with time-varying demand. Sarker et al. (1997) presented an order-level lot size inventory model with inventory-level dependent demand and deterioration. Chen and Wang (1996), Roy and Maiti (1997), Yao et al. (2000) and Chang (2004) have extended the well-known inventory models to fuzzy environment. Kao and Hsu (2002) presented lot size-reorder point inventory model with fuzzy demands. Wang (2002) studied an inventory replenishment policy for deteriorating items with shortages and partial backlogging. Chandra et al (2012) presented fuzzy inventory model for deteriorating items with time-varying demand and shortages. Jaggi et al. (2012) proposed a fuzzy inventory model for deteriorating items with time-varying demand and shortages.

The decaying inventory problem was first presented by Ghare and Shadar (1963) whom developed EOQ model with constant rate of decay, the same was extended by Covert and Philip (1973) for a variable of deterioration. Hollier and Mak (1983) presented inventory replenishment policies for deteriorating items in a declining market. Gen et al. (1997) considered classical inventory model with Triangular fuzzy number. H. Wee (1995) proposed a deterministic lot-size inventory model for deteriorating items with shortages and a declining market. Yao and Lee (1998) considered an economic production quantity model in the fuzzy sense. Chang et al (2003) developed an EOQ model for deteriorating items under supplier credits linked to ordering quantity. Moon et al. (2005) derived an economic order quantity (EOQ) models for ameliorating/deteriorating items under inflation and time discounting. Liang and Zhou (2011) studied a two warehouse inventory model for deteriorating items under conditionally permissible delay in Payment. Alamri and Balkhi (2007) considered the effects of learning and for getting on the optimal production lot size for deteriorating items with time varying demand and deterioration rates. Mahata and Mahata (2011) presented Analysis of a fuzzy economic order quantity model for deteriorating items under retailer partial trade credit financing in a supply chain. R Uthaykumar and M Valliathal (2011) provided fuzzy economic production quantity model for weibull deteriorating items with ramp type of demand.

Fuzzy set theory was demonstrated by Zadeh in 1965. Kauffmann and Gupta (1991) provided an introduction to fuzzy arithmetic operation and Zimmermann (1996) discussed the concept of the fuzzy set theory and its applications. Park (1987) applied the fuzzy set concepts to economic order quantity (EOQ) formula by representing the inventory carrying cost with a fuzzy number and solved the economic order quantity (EOQ) model using fuzzy number operations based on the extension principle. Vujosevic et al. (1996) presented an EOQ Formula when inventory cost is fuzzy. They used trapezoidal fuzzy number (TrFN) to fuzzify the order cost in the total cost of the inventory model without backorder, and got fuzzy total cost. Yao and Lee (1998) considered an economic production quantity model in the fuzzy sense. Lin and Yao (2000) presented the optimal solution for fuzzy case of economic production for production inventory model. Kao and Hsu (2002) presented a single-period inventory model with fuzzy demand. Hsieh (2002) proposed some production inventory models in fuzzy sense. Sujit Kumar De, P.K. Kundu and A. Goswami (2003) presented an economic production quantity inventory model involving fuzzy demand rate. J.K. Syde and L.A. Aziz (2007) applied sign distance method to fuzzy inventory model without shortage. D. Datta and Pravin Kumar published several paper of fuzzy inventory with or without shortage. S. Islam, T.K. Roy (2006) presented a fuzzy EPQ model with flexibility and reliability consideration and demand depended unit Production cost under a space constraint. Roy and Maity (1998) analyzed a multiple objective inventory model for deteriorating items with stock depended demand under two restrictions in fuzzy environment. De and Rawat (2011) derived an EOQ model with-out shortage cost by using triangular fuzzy number. Jana, Das and Maiti (2014) presented multi-item partial backlogging inventory models over random planning horizon in random fuzzy environment.

Since late 1960's, Geometric Programming (GP) used in various field (like OR, Engineering science etc.). Geometric Programming (GP) is one of the effective methods to solve a particular type of non-linear programming (NLP) problem. The theory of Geometric Programming (GP) first emerged in 1961 by Duffin and Zener. The first publication on GP was published by Duffin and Zener on (1967). There are many references on applications and methods of GP in the survey paper by Ecker et al. (1984). They describe GP with positive or zero degree of difficulty. But there may be some problems on GP with negative degree of difficulty. Sinha et al. (1989) proposed it theoretically. Abot-El-Ata (1997) and his group applied modified form of GP in inventory models. Jung and Klein (2001) presented optimal inventory policies under decreasing cost functions via geometric programming. Mandal et al

(2006) proposed an inventory model of deteriorating items with a constraint via a geometric programming approach. A solution method of posynomial geometric programming with interval exponents and coefficients was developed by Liu (2008). Kotb et al. (2011) presented Multi-item EOQ model with both demand-depended unit cost and varying lead time via geometric programming (GP). S.Islam and T.K. Roy (2005) has been proposed modified geometric programming problem. Samir Dey and Tapan Kumar Roy (2015) presented optimum shape design of structural model with imprecise coefficient by parametric geometric programming (GP).

Carlsson and Fuller (2001) developed on Possibilistic Mean Value and Variance of Fuzzy Numbers. Hong and Kim (2004) treated a note on weighted possibilistic mean. Zhang et al. (2006) analyzed a portfolio selection method based on possibility theory. Appadoo, Bhatt and Bector (2008) discussed the application of possibility theory to investment decisions. Paseka, Appadoo and Thavaneswaran (2011) discussed about Possibilistic moment generating functions. Appadoo, Bector and Bhatt (2012) presented fuzzy EOQ model using possibilistic approach. Appadoo, Gajpal and Bhatti (2014) presented a gaussian fuzzy inventory EOQ model subject to inaccuracies in model parameters.

The term "signomial" was introduced by Richard J. Duffin and Elmor L. Peterson in their seminal joint work on general algebraic optimization, published in the late 1960s and early 1970s. A recent introductory exposition is optimization problems. Although non-linear optimization problems with constraints and /or objectives defined by signomials are normally harder to solve than those defined by only posynimials (because unlike posynomals, signomials are not guaranteed to be globally convex). A signomial optimization problem often provides a much more accurate mathematical representation of real-world nonlinear optimization problems. Initially Passy and Wilde (1967), and Blau and Wilde (1969) generalized some of the prototype concepts and theorems in order to treat signomial programs. In other work that general type of signomial programming has been done by Charnes and Cooper (1988), who proposed methods for approximating signomial programs with prototype geometric programs.

1.12   Objective of the Thesis

The basic objective of an inventory model is to reduce total average cost, i.e., maximum profit and also ensuring that the production process does not suffer at the same time. In this thesis, we have developed some inventory models for finding total annual average cost and optimal ordering quantity in crisp and fuzzy environment. In the past researchers assumed the parameters involved in an inventory model such as the demand, holding cost, se-up cost, deteriorating cost, etc. as a crisp values or random variables. But in reality some uncertainty occurs. The demand and cost of the items change from day to day. Again, calculating these variables by the probability distribution is very hard due to lack of historical data. The cost parameters generally estimated based on previous experiment and managerial judgment. Therefore, in real life situation fuzzy set theory is more realistic than crisp set theory or traditional probability theory. Here we have used triangular fuzzy number (TFN) also used fuzzy grade-mean integration representation (GMIR) method. Optimal solution (OS) is obtained here by applying non-linear programming techniques (GP, SGP, Khun –Tucker necessary conditions) for a single item inventory model. But it is observed that the geometric programming (GP) technique minimizes the total annual average cost more than any usual non-linear technique. It is also analyzed that when fuzzification of the given inventory model is done the total average cost is minimized by satisfying the constraints under necessary conditions.

1.13   Organization of the Thesis

The main aim of the thesis to developed some concepts of inventory models approach to decision making in crisp and fuzzy environment. The thesis is organized into eleven chapters. First chapter is introduction, second chapter is methodology, eleventh chapter is conclusion & future work and the rest eight chapters are according to my eight research papers.

In the first chapter, we have studied a brief introduction about operations research (OR), inventory management (IM) and fuzzy set theory. Literature review and organization of thesis are also presented here.

In the second chapter, we have discussed some mathematical concepts (methodology) and techniques which are used in this Thesis for solving inventory models in crisp and fuzzy environment.

In the third chapter, an economic order quantity (EOQ) model with cost of interest, time dependent holding cost, without shortages is formulated and solved. In most real world situation, the objective and constraint function of the decision makers are imprecise in nature. Hence the coefficients are imposed here in fuzzy environment. Here the model is solved by geometric programming (GP), non-linear programming (NLP), fuzzy geometric programming (FGP) and fuzzy non linear programming (FNLP) respectively. At last we have given a numerical example and solved by various methods and, seen that fuzzy geometric programming (FGP) is given better result than any other methods.

In the fourth chapter, we have analyzed fuzzy inventory model for deterioration item with constant demand. Shortages are allowed under fully backlogged. Fixed cost, deterioration cost, shortages cost, holding cost are the cost considered in this model. In this model we have considered a special condition that the demand falls to zero in a time interval $(t_0 \leq t \leq t_e)$ for an unexpected situation (flood, strike, earthquake, etc.) Fuzziness is applying by allowing the cost components (holding cost, deterioration cost, shortages cost, etc.). In fuzzy environment we have considered all required parameter to be triangular fuzzy number (TFN). Here we have used nearest interval approximation (NIA) method to convert a triangular fuzzy number (TFN) to an interval number. In this chapter, we have transformed this interval number to a parametric interval-valued functional form. Several numerical examples are given to verify optimal solutions. The purpose of the model is to minimize total average cost function.

In the fifth chapter, an inventory model with unit production cost, time depended holding cost, without shortages is formulated and solved. We have considered here a single objective inventory model. In most real world situation, the objective and constraint function of the decision makers are imprecise in nature, hence the coefficients are imposed here in fuzzy environment. Geometric programming (GP) provides a powerful tool for solving a variety of optimization problem. Here we have used nearest interval approximation (NIA) method to convert a triangular fuzzy number (TFN) to an interval number thereafter transformed this interval number to a parametric interval-valued functional form and solve the parametric problem by geometric programming (GP) technique. Here two necessary theorems have been derived. Numerical example is given to illustrate the model through this parametric geometric programming (PGP) technique.

In the sixth chapter, an economic order quantity (EOQ) model with unit production cost, time dependent holding cost, without shortages is formulated and solved. In most real world situation, the objective and constraint function of the decision makers are imprecise in nature, so the coefficients are imposed here in fuzzy environment. The problem is then solved using fuzzy max-min geometric programming technique and fuzzy parametric geometric programming technique respectively. Sensitivity analysis is also presented here.

In the seventh chapter, we have analyzed fuzzy inventory system for deterioration item with time depended demand. Shortages are allowed under fully backlogged. Fixed cost, deterioration cost, shortages cost, holding cost are the cost considered in this model. Fuzziness is applying by allowing the cost components (fixed cost, deterioration, shortage cost, holding cost, etc.). In fuzzy environment we have considered all required input parameter to be triangular fuzzy number (TFN). One numerical solution of the model is obtained to verify optimal solution (OS). The purpose of the model is to minimize total average cost function.

In the eighth chapter, we have proposed a fuzzy inventory model with constant demand without shortages in a fuzzy environment. Here we have used Bell shaped fuzzy membership function (MF). In this chapter we have developed the concepts of possibility theory and possibilistic moment generating functions and some statistical concept as mean, variance, standard deviation on this economic order quantity (EOQ) model. Also some necessary theorems have been derived here. Finally, these are illustrated by numerical examples and applications.

In the ninth chapter, we have developed a fuzzy inventory model with shortages under fully backlogging with constant demand in a fuzzy environment. Here we have used Bell shaped fuzzy membership function. In this chapter we have developed the concept of possibility theory and possibilistic moment generating function. Three type of possibilistic mean values as lower possibilistic$(E_L(A))$, upper possibilistic$(E_R(A))$ and crisp possibilistic (E(A)) of total average cost function is developed here. Also some necessary theorems have been derived. Finally, these are illustrated by numerical examples and applications.

In the tenth chapter, a fuzzy economic order quantity (EOQ) model with shortages under fully backlogging and constant demand is formulated and solved. Here the model is solved by fuzzy signomial geometric programming (FSGP) technique. Fuzzy signomial geometric programming (FSGP) method provides a powerful technique for solving some special non-linear problems. Here we have proposed a new idea, that is fuzzy modified signomial geometric programming (FMSGP) technique and some necessary theorems have been derived. Finally, these are illustrated by some numerical examples and applications.

The last chapter, i.e., **eleventh chapter** is conclusions & future work. This chapter gives an overview of the complete research project and its conclusions and outlines further directions to extend and advance the current research.

Chapter 2

Methodology

In this chapter we have analyzed some mathematical concepts and methods which are used for solving inventory models in crisp and fuzzy environment. This chapter contains four sections. First section is unconstrained programming technique, second is constrained programming technique, third is fuzzy grade-mean integration representation (GMIR) method and fourth is Khun-Tucker necessary conditions, defined as follows:

i) **Unconstrained Programming**

    a) *Unconstrained Geometric Programming (UGP)*

    b) *Unconstrained Modified Geometric Programming (UMGP)*

    c) *Unconstrained Signomial Geometric Programming (USGP)*

    d) *Unconstrained Modified Signomial Geometric Programming (UMSGP)*

ii) **Constrained Programming**

    a) *Constrained Geometric Programming (CGP)*

    b) *Constrained Modified Geometric Programming (CMGP)*

    c) *Constrained Signomial Geometric Programming (CSGP)*

    d) *Constrained Modified Signomial Geometric Programming (CMSGP)*

iii) **Grade-Mean Integration Representation (GMIR) Method**

iv) **Khun-Tuckar Necessary Conditions**

2.1 Unconstrained Program

2.1.1 Unconstrained geometric programming

Primal Problem

Primal Geometric Programming (PGP) problem is:

$$\text{Minimize} \quad g(t) = \sum_{k=1}^{T_0} c_k \prod_{j=1}^{n} t_j^{\alpha_{kj}} \tag{2.1}$$

Subject to $t_j > 0$, (j = 1, 2,....,n).

Here c_k (>0), and α_{kj} (j =1,2,....,n; k = 1,2,..,T_0) be any real number and Degrees of Difficulty (DD) of a GP = No. of terms in PGP − (1+ No. of variables in PGP).

Dual Problem

Dual Programming (DP) problem of (2.1) is:

$$Maximize\ v(\delta) = \prod_{k=1}^{T_0} \left(\frac{c_k}{\delta_k} \right)^{\delta_k} \tag{2.2}$$

Subject to

$$\sum_{k=1}^{T_0} \delta_k = 1 \qquad\qquad \text{(Normality condition)}$$

$$\sum_{k=1}^{T_0} \alpha_{kj}\delta_k = 0,\ \text{(j=1,2,....,n)} \qquad \text{(Orthogonality conditions)}$$

$$\delta_k > 0,\ \ \text{(k=1,2,.......,}T_0\text{).} \qquad \text{(Positivity conditions)}$$

Where $\delta = (\delta_1, \delta_2,......, \delta_{T_0})^T$.

Case I: T_0 = n+1, (i.e. DD=0) so DP presents a system of linear equations for the dual variables. A unique solution vector of dual variable exists. In this case no optimization is necessary. There is a single feasible solution of DP.

Case II: T_0 > n+1, (i.e. DD > 0) so the DP presents a system of linear equations for the dual variables. Here the number of linear equations is less than the number of dual variables. More solutions of dual variable vector exist. In order to find an optimal solution of DP, we need to use some algorithmic methods.

Case III: T_0 < n+1, (i.e. DD < 0) so the DP presents a system of linear equations for the dual variables. Here the number of linear equations is greater than the number of dual variables. In this case generally no solution vector exists for the dual variables. However, using Least Square (LS) or Min-Max (MM) method one can get an approximate solution for this system.

Once optimal dual variable vector δ^* is known, the corresponding values of the primal variable vector t is found from the following relations:

$$c_k \prod_{j=1}^{n} t_j^{\alpha_{kj}} = \delta_k^* v^*(\delta^*), \qquad\qquad \text{(k = 1, 2,...,}T_0\text{)} \tag{2.3}$$

Taking logarithms in (6), T_0 log-linear simultaneous equations are transformed as

$$\sum_{j=1}^{n} \alpha_{kj} (\log t_j) = \log\left(\frac{\delta_k^* v^*(\delta^*)}{c_k} \right), \quad (k = 1, 2, \ldots, T_0) \tag{2.4}$$

It is a system of T_0 linear equations in $x_j (= \log t_j)$ for j=1, 2,...,n.

2.1.2 Unconstrained modified geometric programming (UMGP) problem

Primal Program:

Special type of unconstrained PGP problem is

$$Minimize \quad g(t) = \sum_{i=1}^{n} g_i(t_i) = \sum_{i=1}^{n} \sum_{k=1}^{T_0} C_{ik} \prod_{j=1}^{m} t_{ij}^{\alpha_{ikj}} \tag{2.5}$$

Subject to $t_{ij} > 0$, (i =1,2,......,n; j=1,2,......,m.

$$t = (t_{11}, t_{12}, \ldots\ldots, t_{1n})^T, t_i = (t_{i1}, t_{i2}, \ldots\ldots, t_{in})^T, (i = 1,2,\ldots\ldots, n).$$

Where $C_{ik}(> 0)$ and α_{ikj} (i =1,2,...,n; j =1,2,...,m; k =1,2,...,T_0) are real numbers.

Here g(t) is a sum of n separable posynomial functions $g_i(t_i)$ (i =1,2,...,n) of distinct variables.

Dual Program:

Applying the GP techniques to (2.5), the enlarged pre-dual function would be written in the following form

$$Maximize \quad v(\delta) = \prod_{i=1}^{n} \prod_{k=1}^{T_0} \left(\frac{c_{ik}}{\delta_{ik}} \right)^{\delta_{ik}} \tag{2.6}$$

Subject to

$$\sum_{k=1}^{T_0} \delta_{ik} = 1 \qquad\qquad \text{(Normality conditions)}$$

$$\sum_{k=1}^{T_0} \alpha_{ikj} \delta_{ik} = 0 \quad (j=1,2,...,m) \qquad \text{(Orthogonality conditions)}$$

$$\delta_{ik} > 0, \ (k=1,2,\ldots,T_0), i=1,2,\ldots,n. \qquad \text{(Positivity conditions)}$$

Here δ is a decision variable vector of nT_0 components i.e. $(\delta_{11}, \delta_{12}, \ldots, \delta_{ik}, \ldots, \delta_{nT_0})^T$

Case I: $nT_0 \leq nm+n$, so DP presents a system of (mn+n) linear equations with nT_0 variables. So, a unique solution set of dual variables exists.

Case II: $nT_0 > nm+n$, so the DP presents a system of (mn+n) linear equations for the nT_0 dual variables. A solution vector for the dual variables exists. Here no of linear equations is less than the no. of dual variables. So many solutions of dual variable exist.

Case III: $nT_0 < nm+n$, in this case generally no solution of dual variables exists. However, using either the LS or MM method one can get an approximate solution vector for this system.

Theorem 2.1.2(a): If t is a feasible vector for the unconstrained PGP (2.5) and if δ is a feasible vector for the corresponding DP (2.6), then $g(t) \geq n\sqrt[n]{v(\delta)}$ (Primal-Dual inequality).

Proof (Islam and Roy (2005)).

Theorem 2.1.2(b): If $t^* = (t^*_{i1},\ldots\ldots,t^*_{im})$ for i = 1,2,…..,n is a solution to the PGP (2.5) ,then the corresponding DP (2.6) is consistent. Moreover the vector $\delta^* = (\delta^*_{i1},\ldots,\delta^*_{iT_0})$ for $i = 1,2,\ldots\ldots,n$ defined by

$$\delta^*_{ik} = \frac{u_{ik}(t^*)}{g_i(t^*)} , \quad (i = 1, 2,\ldots\ldots, n; k =1,2,\ldots\ldots,T_0).$$

Where $u_{ik}(t^*) = c_{ik}\prod_{j=1}^{m} t_{ij}^{*\alpha_{ikj}}$ is the kth term of $g(t)$ for i-th item is a solution for dual program (DP) and equality holds in the primal-dual inequality i.e. $g(t^*) = n\sqrt[n]{v(\delta^*)}$.

Proof (Islam and Roy (2005))

Now we can get dual variables which optimize $v^*(\delta^*)$ such that

$$c_{ik}\prod_{j=1}^{m} t_{ij}^{\alpha_{ikj}} = \delta_{ik}\sqrt[n]{v(\delta)}, \quad (i =1, 2,\ldots, n; k =1,2,\ldots,T_0) \tag{2.7}$$

Taking logarithms in (15), the linear simultaneous equations are transformed as

$$\sum_{j=1}^{m}\alpha_{ikj}(Logt_{ij}) = Log\left(\frac{\delta_{ik}\sqrt[n]{v(\delta)}}{c_{ik}}\right), \quad (i =1,2,\ldots,n; k=1,2,..,T_0). \tag{2.8}$$

It is a system of linear equations in x_{ij} (= Log t_{ij}) (i =1, 2,..., n; j = 1,2,..m). Since there are more primal variables t_{ij} than the number of terms nT_0, many solutions t_{ij} (i = 1,2,...,n; j = 1,2,..,m) may exist. So, to find the optimal primal variables t_{ij} (i = 1,2,..,n; j = 1,2,..,m), it remains to minimize the primal objective function with respect to reduced $nm - nT_0$ (≠0) variables. When $nm - nT_0 = 0$, primal variables can be determined uniquely from log-linear equations.

2.1.3 Unconstrained signomial GP problem:

Primal program:

A primal unconstrained signomial GP programming problem is of the form

$$\text{Minimize} \quad g_0(x_1, x_2, \ldots \ldots \ldots, x_m) \tag{2.9}$$

$$\text{Subject to} \quad x_j > 0, \ j = 1, 2, \ldots \ldots, m.$$

Where $g_0(x) = \sum_{i=1}^{n} \sigma_i c_i \prod_{j=1}^{m} x_j{}^{a_{ij}}$.

Here c_i is absolute value of coefficient, σ_i is sign of coefficient $(+1 \text{ or } -1)$ and a_{ij} be any real number. It is unconstrained signomial GP problem with the degree of difficulty (DD) $= n - (m + 1)$.

Dual program:

Dual GP problem of the given primal GP problem is

$$\text{Maximize} \quad v(\delta) = \zeta_0 \left[\prod_{i=1}^{n} (\frac{c_i}{\delta_i})^{\sigma_i \delta_i} \right]^{\zeta_0} \tag{2.10}$$

Subject to

$$\sum_{i=1}^{n} \sigma_i \delta_i = \zeta_0,$$

$$\sum_{i=1}^{n} \sigma_i a_{ij} \delta_i = 0, \qquad j = 1, 2, \ldots \ldots \ldots, m,$$

$$\delta_i > 0.$$

Case I: $n > m+1$, (i.e. DD >0) so the DP presents a system of linear equations for the dual variables. Here the number of linear equations is less than the number of dual variables. More solutions of dual variable vector exist. In order to find an optimal solution of DP, we need to use some algorithmic methods.

Case II: $n < m+1$, (i.e. DD <0) so the DP presents a system of linear equations for the dual variables. Here the number of linear equations is greater than the number of dual variables. In this case generally no solution vector exists for the dual variables. However, using Least Square (LS) or Min-Max (MM) method one can get an approximate solution for this system.

Furthermore the primal-dual relation is

$$c_i \prod_{j=1}^{m} x_j^{*a_{ij}} = \zeta_0 \delta^*_i \, v(\delta^*, \lambda^*). \tag{2.11}$$

Taking logarithms in (3.3), T_0 log-linear simultaneous equations are transformed as

$$\sum_{j=1}^{n} \alpha_{ij} (\log x_j) = \log\left(\frac{\zeta_0 \delta^*_i v^*(\delta^*, \lambda^*)}{c_i} \right), \quad (i = 1, 2, \ldots, n). \tag{2.12}$$

It is a system of "n" linear equations in $t_j (= \log x_j)$ for $j = 1, 2, \ldots, m$.

Note: A Weak Duality theorem would say that

$$g_0(x) \geq v(\delta)$$

For any primal-feasible x and dual-feasible δ but this is not true of the pseudo-dual signomial GP problem.

Corollary: When the value of σ_i is 1, then a signomial geometric programming problem transform to ordinary geometric programming problem.

Theorem 2.1.3(a): When σ_i is 1, then $g_0(x) \geq v(\delta)$ (Primal- Dual Inequality).

Proof

The expression for $g_0(x)$ can be written as

$$g_0(x) = \sum_{i=1}^{n} \delta_k \left(\frac{c_i \prod_{j=1}^{m} x_j^{\alpha_{kj}}}{\delta_k} \right).$$

Here the weights are $\delta_1, \delta_2, \ldots\ldots\ldots, \delta_n$ and positive terms are $\dfrac{c_1 \prod_{j=1}^{m} x_j^{\alpha_{1j}}}{\delta_1}, \dfrac{c_2 \prod_{j=1}^{m} x_j^{\alpha_{2j}}}{\delta_2}, \ldots\ldots\ldots,$

$\dfrac{c_n \prod_{j=1}^{m} x_j^{\alpha_{nj}}}{\delta_n}.$

Now applying A.M.-.G.M inequality, we get

$$\left(\frac{c_1 \prod_{j=1}^{m} x_j^{\alpha_{1j}} + c_2 \prod_{j=1}^{m} x_j^{\alpha_{2j}} + \ldots + c_n \prod_{j=1}^{m} x_j^{\alpha_{nj}}}{(\delta_1 + \delta_2 + \cdots + \delta_n)} \right)^{(\delta_{01} + \delta_{02} + \cdots + \delta_n)}$$

$$\geq \left((\frac{c_1 \prod_{j=1}^{m} x_j{}^{\alpha_{1j}}}{\delta_1})^{\delta_1} (\frac{c_2 \prod_{j=1}^{m} x_j{}^{\alpha_{2j}}}{\delta_2})^{\delta_2} \dots (\frac{c_n \prod_{j=1}^{m} x_j{}^{\alpha_{nj}}}{\delta_n})^{\delta_n} \right)$$

Or $\quad \left(\dfrac{g_0(x)}{\sum_{i=1}^{n} \delta_i} \right)^{\sum_{i=1}^{n} \delta_i} \geq \prod_{i=1}^{n} \left(\dfrac{c_i \prod_{j=1}^{m} x_j{}^{\alpha_{nj}}}{\delta_i} \right)^{\delta_i}$ $\qquad [as \; \sum_{i=1}^{n} \delta_k = 1]$

Or $\quad g_0(x) \geq \left(\dfrac{c_i}{\delta_{0k}} \right)^{\sum_{i=1}^{n} \delta_i} \prod_{j=1}^{m} x_j{}^{\sum_{i=1}^{n} \alpha_{ij} \delta_i}$

Or $\quad g_0(x) \geq \prod_{i=1}^{n} \left(\dfrac{c_i}{\delta_i} \right)^{\delta_i} \prod_{j=1}^{m} x_j{}^{\sum_{i=1}^{n} \alpha_{ij} \delta_i}$

$\qquad\qquad = \prod_{i=1}^{n} \left(\dfrac{c_i}{\delta_i} \right)^{\delta_i}$ $\qquad\qquad [as \; \sum_{k=1}^{T_0} \alpha_{0kj} \delta_{ok} = 0]$

$\qquad\qquad = v(\delta)$

i.e., $\quad g_0(x) \geq v(\delta)$.

2.1.4 Unconstrained modified signomial geometric programming (UMSGP) problem

Primal program:

A primal unconstrained modified signomial GP programming problem is of the form

$\qquad$ Minimize $\quad g_0(x_{lj})$

$\qquad$ Subject to $\quad x_{lj} > 0$, j = 1, 2,……,m.

Where $g_0(x) = \sum_{l=1}^{n} \sum_{i=1}^{k} \sigma_{li} c_{li} \prod_{j=1}^{m} x_{lj}{}^{a_{lij}}$. $\qquad\qquad$ (2.13)

Here c_{li} is absolute value of coefficient, σ_{li} is sign of coefficients (+1 or − 1) and a_{lij} be any real number. It is unconstrained signomial MGP problem with the degree of difficulty (DD) $= nk - (nm + n)$.

Dual program:

Dual GP problem of the given primal GP problem is

$\qquad$ Maximize $\zeta_0 \left[\prod_{l=1}^{n} \prod_{i=1}^{k} (\dfrac{c_{li}}{\delta_{li}})^{\sigma_{li} \delta_{li}} \right]^{\zeta_0}$ $\qquad\qquad$ (2.14)

$\qquad$ Subject to

$$\sum_{i=1}^{k} \sigma_{li} \delta_{li} = \zeta_0, \qquad (l = 1,2,\dots\dots\dots,n).$$

$$\sum_{i=1}^{k} \sigma_{li} a_{lij} \delta_{li} = 0, \qquad (l = 1,2,\dots\dots\dots,n; j = 1,2,\dots\dots\dots,m).$$

$$\delta_{li} > 0,$$

Case I: $nk \geq nm+n$, (i.e. DD>0) so the DP presents a system of linear equations for the dual variables. Here the number of linear equations is less than the number of dual variables. More solutions of dual variable vector exist. In order to find an optimal solution of DP, we need to use some algorithmic methods.

Case II: $nk < nm+n$, (i.e. DD <0) so the DP presents a system of linear equations for the dual variables. Here the number of linear equations is greater than the number of dual variables. In this case generally no solution vector exists for the dual variables. However, using Least Square (LS) or Min-Max (MM) method one can get an approximate solution for this system.

Furthermore the primal-dual relation is

$$c_{li} \prod_{j=1}^{m} x_{lj}{}^{a_{lij}} = \zeta_0 \delta^*{}_{li} \sqrt[n]{v(\delta^*)}., \ (l = 1,2,\dots\dots\dots,k; i = 1,2,\dots\dots\dots,n). \qquad (2.15)$$

Note 2: A Weak Duality theorem would say that

$$g_0(x_{lj}) \geq n\sqrt[n]{v(\delta)}.$$

For any primal-feasible x and dual-feasible δ but this is not true of the pseudo-dual signomial GP problem.

Corollary: When the values of σ_{li} is 1, then a modified signomial geometric programming problem transform to ordinary modified geometric programming problem.

Theorem 2.1.4(a): When σ_i is 1, then $g_0(x_{ij}) \geq n\sqrt[n]{v(\delta)}$ (Primal- Dual Inequality).

Proof

The expression for $g_0(x_{ij})$ can be written as

$$g_0(x_{ij}) = \sum_{l=1}^{n} \sum_{i=1}^{k} \delta_{ik} \left(\frac{c_{li} \prod_{j=1}^{m} x_{ij}{}^{\alpha_{lij}}}{\delta_{ik}} \right).$$

Here the weights are $\delta_{l1}, \delta_{l2}, \dots\dots\dots, \delta_{lk}$ and positive terms are

$$\frac{c_{l1} \prod_{j=1}^{m} x_j{}^{\alpha_{l1j}}}{\delta_{l1}}, \frac{c_{l2} \prod_{j=1}^{m} x_j{}^{\alpha_{l2j}}}{\delta_{l2}}, \dots\dots\dots, \frac{c_{lk} \prod_{j=1}^{m} x_j{}^{\alpha_{lnj}}}{\delta_{lk}}.$$

Now applying A.M.-.G.M inequality, we get

$$\left(\frac{\sum_{l=1}^{n}(c_{l1}\prod_{j=1}^{m}x_{ij}{}^{\alpha_{l1j}}+c_{l2}\prod_{j=1}^{m}x_{ij}{}^{\alpha_{l2j}}+\cdots+c_{ln}\prod_{j=1}^{m}x_{ij}{}^{\alpha_{lkj}})}{\sum_{l=1}^{n}(\delta_{l1}+\delta_{l2}+\cdots+\delta_{lk})}\right)^{\sum_{i=1}^{n}(\delta_{l1}+\delta_{l2}+\cdots+\delta_{lk})}$$

$$\geq \sum_{l=1}^{n}\left((\frac{c_{l1}\prod_{j=1}^{m}x_{ij}{}^{\alpha_{i1j}}}{\delta_{l1}})^{\delta_{l1}}(\frac{c_{l2}\prod_{j=1}^{m}x_{ij}{}^{\alpha_{l2j}}}{\delta_{l2}})^{\delta_{l2}}\cdots(\frac{c_{lk}\prod_{j=1}^{m}x_{ij}{}^{\alpha_{lkj}}}{\delta_{lk}})^{\delta_{lk}}\right)$$

Or $\quad\left(\dfrac{g_0(x_{ij})}{\sum_{l=1}^{n}\sum_{i=1}^{k}\delta_{li}}\right)^{\sum_{l=1}^{n}\sum_{i=1}^{k}\delta_{li}} \geq \prod_{l=1}^{n}\prod_{i=1}^{k}\left(\dfrac{c_{li}\prod_{j=1}^{m}x_{ij}{}^{\alpha_{lij}}}{\delta_{ik}}\right)^{\delta_{ik}}$

Or $\quad\left(\dfrac{g_0(x_{ij})}{n}\right)^{n} \geq \prod_{l=1}^{n}\left(\dfrac{c_{li}}{\delta_{lk}}\right)^{\sum_{l=1}^{k}\delta_{lk}}\prod_{j=1}^{m}x_{ij}{}^{\sum_{i=1}^{k}\alpha_{lij}\delta_{li}} \qquad\qquad [as\ \sum_{i=1}^{k}\delta_{li}=1]$

$$= \prod_{l=1}^{n}\prod_{i=1}^{k}\left(\dfrac{c_{li}}{\delta_{li}}\right)^{\delta_{li}}\prod_{j=1}^{m}x_{ij}{}^{\sum_{i=1}^{k}\alpha_{lij}\delta_{li}}$$

Or $\quad\left(\dfrac{g_0(x_{ij})}{n}\right)^{n} \geq \prod_{l=1}^{n}\prod_{i=1}^{k}\left(\dfrac{c_{li}}{\delta_{li}}\right)^{\delta_{li}} \qquad\qquad\qquad [as\ \sum_{i=1}^{k}\alpha_{lij}\delta_{li}=0]$

$$= v(\delta)$$

i.e., $\quad g_0(x_{ij}) \geq n\sqrt[n]{v(\delta)}.$

2.2   Constrained Programming Problem

2.2.1 Constrained geometric programming (CGP) problem

Primal Geometric Programming (PGP):

$$\text{Minimize}\quad g_0(t) = \sum_{k=1}^{T_0}c_{0k}\prod_{j=1}^{m}t_j^{\alpha_{0kj}} \qquad\qquad (2.16)$$

$$\text{Subject to}\quad g_r(t) = \sum_{k=1+T_{r-1}}^{T_r}c_{rk}\prod_{j=1}^{m}t_j^{\alpha_{rkj}} \leq 1$$

$$t_j > 0,\ j = 1,2,\ldots,m.$$

Where $c_{rk}(>0)$ and α_{rkj} (k = 1,2,..,1+T_{r-1},..,T_r;r = 0,1,2,..,l;j = 1,2,...,m) are real numbers.

It is a constrained posynomial geometric programming (PGP) problem. The number of terms in each posynomial constraint function varies and it is denoted by T_r for each r = 0, 1, 2,...,l. Let T=T_0+T_1+T_2+.........+T_l, be the total number of terms in the primal program. Then the Degree of Difficulty (DD) = T− (m+1).

Dual Program:

The dual programming of (2.16) is as follows:

$$\text{Maximize } v(\delta) = \prod_{r=0}^{l}\prod_{k=1}^{T_r}\left(\frac{c_{rk}}{\delta_{rk}}\right)^{\delta_{rk}}\left(\sum_{s=1+T_{r-1}}^{T_r}\delta_{rs}\right)^{\delta_{rk}} \qquad (2.17)$$

Subject to

$$\sum_{k=1}^{T_0}\delta_{0k} = 1, \qquad \text{(Normality condition)}$$

$$\sum_{r=0}^{l}\sum_{k=1}^{T_r}\alpha_{rkj}\delta_{rk} = 0, \quad (j=1,2,...,m) \qquad \text{(Orthogonality conditions)}$$

$$\delta_{rk} > 0, \quad (r=0,1,2,...,l; \ k=1,2,..,T_r). \quad \text{(Positivity conditions)}$$

Case I: $T_0 = m+1$, So DP presents a system of linear equations for the dual variables. A unique solution vector of dual variable exists.

Case II: $T_0 > m+1$, So the DP presents a system of linear equations for the dual variables. Here the number of linear equations is less than the number of dual variables. More solutions of dual variable vector exist.

Case III: $T_0 < m+1$, So the DP presents a system of linear equations for the dual variables. Here the number of linear equations is greater than the number of dual variables. In this case generally no solution vector exists for the dual variables. However, using Least Square (LS) or Min-Max (MM) method one can get an approximate solution for this system.

The solution procedure of this GP problem is same as the unconstrained GP problem.

2.2.2 Constrained modified geometric programming (CMGP) problem

Primal problem:

$$\text{Minimize} \quad g_0(t) = \sum_{i=1}^{n}g_{i0}(t_i) = \sum_{i=1}^{n}\sum_{k=1}^{T_0}C_{0ik}\prod_{j=1}^{m}t_{ij}^{\alpha_{0ikj}} \qquad (2.18)$$

$$\text{Subject to} \quad \sum_{i=1}^{n}g_{ir}(t_i) = \sum_{i=1}^{n}\sum_{k=T_{r-1}+1}^{T_r}c_{rik}\prod_{j=1}^{m}t_{ij}^{\alpha_{rikj}} \leq 1,$$

$$t_{ij} > 0 \ (i = 1,2,...,n; \ j = 1,2,...,m).$$

Where, $C_{ik}(>0)$ and α_{rikj} ($i=1,2,...,n; \ j = 1,2,...,m; \ r=0,1,2,...,l; \ k = 1,2,...,T_{r-1}+1,...,T_r$) are real number.

Dual Program:

Applying the GP techniques to equation (2.18), the enlarged pre-dual function could be written in the following form

$$\text{Maximize } v(\delta) = \prod_{i=1}^{n} \prod_{k=1}^{T_r} \left(\frac{c_{ik}}{\delta_{ik}} \right)^{\delta_{ik}} \left(\sum_{s=1}^{n} \delta_{sk} \right)^{\delta_{ik}}$$

Subject to   (2.19)

$$\sum_{k=1}^{T_0} \delta_{ik} = 1 \qquad\qquad \text{(Normality conditions)}$$

$$\sum_{k=1}^{T_0} \alpha_{0ikj} \delta_{ik} + \sum_{r=1}^{l} \sum_{k=T_{r-1}+1}^{T_r} \alpha_{rikj} \delta_{ik} = 0 \;\; (j=1,2,...,m) \quad \text{(Orthogonally conditions)}$$

$$\delta_{ik} > 0, \qquad\qquad\qquad \text{(Positivity conditions)}$$

for i =1,2,..,n; where, $\delta = (\delta_{11}, \delta_{12},...., \delta_{ik},...., \delta_{nT_0})^{\text{T}}$.

Case I: $nT_0 = nm+n$, so DP presents a system of $(mn+n)$ linear equations with nT_0 variables. So, a unique solution set of dual variables exists.

Case II: $nT_0 > nm+n$, so the DP presents a system of $(n+mn)$ linear equations for the nT_0 dual variables. A solution vector for the dual variables exists. Here number of linear equations is less than the number of dual variables. So many solutions of dual variables exist.

Case III: $nT_0 < nm+n$, in this case generally no solution of dual variables exists. However, using either the LS or MM method one can get an approximate solution vector for this system.

2.2.3   Constrained signomial GP problem:

Primal program

A primal signomial GP programming problem is of the form

$$\text{Minimize} \quad g_0(x_1, x_2, ..., x_m)$$

$$\text{Subject to} \quad g_k(x_1, x_2, ..., x_m) \leq \zeta_k, \;\; k = 1, 2, ...,p$$

$$x_j > 0, \; j = 1, 2,......,m.$$

Where $g_k(x) = \sum_{i=1}^{k} \sigma_i c_i \prod_{j=1}^{m} x_j{}^{a_{ij}}$, and $\zeta_k = \pm 1$.   (2.20)

Dual program

Dual GP problem of the given primal GP problem is

Maximize $\zeta_0 \left[\prod_{i=1}^{n} (\frac{c_i}{\delta_i})^{\sigma_i \delta_i} \prod_{k=1}^{p} \lambda_k{}^{\zeta_k \lambda_k} \right]^{\zeta_0}$ (2.21)

Subject to

$$\sum_{i=1}^{k} \sigma_i \delta_i = \zeta_k \lambda_k, \qquad k = 0,1, \ldots \ldots \ldots , p.$$

$$\sum_{i=1}^{k} \sigma_i a_{ij} \delta_i = 0, \qquad j = 1,2, \ldots \ldots \ldots , m.$$

$$\delta_i > 0, \ \lambda_0 = 1,$$

Case I: n≥ m+1, (i.e. DD >0), the dual signomial program presents a system of linear equations for the dual variables. A solution vector exists for the dual signomial variables.

Case II: n< m+1, (i.e. DD <0), In this case generally no solution vector exists for the dual signomial variables. But using Least Square (LS) or Min-Max (MM) method one can get an approximate solution for this system.

Furthermore the primal-dual relation is

$$c_i \prod_{j=1}^{m} x_j{}^{*a_{ij}} = \zeta_0 \delta^*{}_i \ v(\delta^*, \lambda^*) \quad \text{for i = 0,}$$

And $c_i \prod_{j=1}^{m} x_j{}^{*a_{ij}} = \delta^*{}_i / \lambda_k{}^* \quad \text{for i} \in [k], \text{i} \geq 1.$ (2.22)

2.2.4 Constrained modified signomial GP programming

Primal program

A primal modified signomial GP programming problem is of the form

Minimize $g_0(x_{lj})$

Subject to $g_{lk}(x) \leq \zeta_{lk}, \quad k = 1, 2, \ldots \ldots , p$

$x_{lj} > 0, \ j = 1, 2, \ldots \ldots , m.$

Where $g_{lk}(x) = \sum_{l=1}^{n} \sum_{i=1}^{k} \sigma_{li} c_{li} \prod_{j=1}^{m} x_{lj}{}^{a_{lij}}$, $\zeta_{00} = 0$ and $\zeta_{lk} = \pm 1$. (2.23)

Dual program

Dual GP problem of the given primal GP problem is

Maximize $\zeta_0 \left[\prod_{l=1}^n \prod_{i=1}^k (\frac{c_{li}}{\delta_{li}})^{\sigma_{li}\delta_{li}} \prod_{k=1}^p \lambda_k{}^{\zeta_k \lambda_k} \right]^{\zeta_0}$ $\hspace{2cm}$ (2.24)

Subject to

$$\sum_{i=1}^k \sigma_{li}\delta_{li} = \zeta_{00}, \hspace{2cm} l = 0,1,\ldots\ldots\ldots,n,$$

$$\sum_{i=1}^k \sigma_{li} a_{ij} \delta_i = 0, \hspace{2cm} j = 1,2,\ldots\ldots\ldots,m,$$

$$\sum_{i=1}^k \sigma_{li}\delta_{li} = \zeta_k \lambda_k, \hspace{2cm} k = 0,1,\ldots\ldots\ldots,p,$$

$$\delta_{li} > 0, \ \lambda_0 = 1.$$

Case I: nk$\geq$ $nm+n$, (i.e. DD > 0), the dual signomial program presents a system of linear equations for the dual variables. A solution vector exists for the dual signomial variables.

Case II: nk< nm+n, (i.e. DD < 0), In this case generally no solution vector exists for the dual signomial variables. But using Least Square (LS) or Min-Max (MM) method one can get an approximate solution for this system.

Furthermore the primal-dual relation

$c_{li} \prod_{j=1}^m x_{lj}{}^{a_{lij}} = \zeta_0 \delta^*{}_{li} \sqrt[n]{v(\delta^*,\lambda^*)}.,$ $(l = 1,2,\ldots\ldots\ldots,k; i = 1,2,\ldots\ldots\ldots,n)$ for i = 0, (2.25)

And $\ c_{li} \prod_{j=1}^m x_{lj}{}^{a_{lij}} = \delta^*{}_{li} / \lambda_{lk}{}^*$ for i $\epsilon[k]$, i $\geq$ 1.

2.3 Grade-Mean Integration Represent (GMIR) Method

For a decision making, the fuzzy results are to be converted into crisp values. The method, converted fuzzy results into crisp results is known as defuzzification. In 1998 grade-mean integration represent method introduced by Chen and Hsieh for defuzzification. The grade-mean integration represents (GMIR) method as follows;
Considered a triangular fuzzy number $\tilde{A} = (a, b, c)$ defined as;

$$\mu_{\tilde{A}}(X) = \begin{cases} L(x) = \frac{x-a}{b-a} \ , & a \leq x \leq b \\ R(x) = \frac{c-x}{c-b} \ , & b \leq x \leq c \\ \quad 0 \qquad , & otherwise \end{cases} \hspace{1cm} (2.26)$$

Suppose L^{-1} and R^{-1} are inverse function of L and R respectively. Then the grade mean h-level value of $\tilde{A}$ is $\frac{h(L^{-1}(h)+R^{-1}(h))}{2}$. Then the grade mean integration representation of fuzzy number $\tilde{A}$ can be represented as

$$P(\tilde{A}) = \frac{\int_0^1 \left(\frac{h(L^{-1}(h)+R^{-1}(h))}{2}\right)dh}{\int_0^1 h\,dh} = \int_0^1 h(L^{-1}(h) + R^{-1}(h))dh.$$

For triangular fuzzy number $L^{-1}(h) = a + (b - a)h$ and $R^{-1}(h) = c - (c - b)h$. Then the grade mean integration representation is

$$P(\tilde{A}) = \int_0^1 h(L^{-1}(h) + R^{-1}(h))dh.$$

$$= \int_0^1 h(a + (b - a)h + c - (c - b)h)dh$$

$$= \frac{a+4b+c}{6}. \tag{2.27}$$

In case of trapezoidal fuzzy number (TrFN) $\tilde{A} = (a, b, c, d)$,

$$P(\tilde{A}) = \frac{a+2b+2c+d}{6}. \tag{2.28}$$

In case of pentagonal fuzzy number (PFN) $\tilde{A} = (a, b, c, d, e)$,

$$P(\tilde{A}) = \frac{a+3b+4c+3d+e}{12}. \tag{2.29}$$

2.4   Kuhn-Tucker Necessary Conditions

$$\text{Maximize} \quad f(x), \; x = (x_1, x_2,\ldots\ldots,x_n) \tag{2.30}$$

$$\text{Subject to} \quad g_i(x) \le b_i, \quad i=1,2,\ldots\ldots,m.$$

Including the non-negative constraints $x \ge 0$, the necessary conditions for a local maxima at x are

i) $\dfrac{\partial L(\bar{X},\bar{\lambda},\bar{s})}{\partial x_j} = 0, \quad j=1,2,\ldots\ldots,n.$

ii) $\bar{\lambda}_i[g_i(\bar{x}) - b_i] = 0,$

iii) $g_i(\bar{x}) \le b_i,$

iv) $\bar{\lambda}_i \ge 0, \quad i =1, 2,\ldots\ldots,m.$

Where $L(\bar{x}, \bar{\lambda}, \bar{s})$ is Lagrange function L defined by:

$L(x_1, x_2,\ldots\ldots, x_n; \lambda_1, \lambda_2,\ldots\ldots, \lambda_m) = f(x_1, x_2,\ldots\ldots, x_n) + \lambda_1 g_1(x_1, x_2,\ldots\ldots, x_n) + \lambda_2 g_2(x_1, x_2,\ldots\ldots, x_n) +\ldots\ldots\ldots+ \lambda_m g_m(x_1, x_2,\ldots\ldots,x_n).$ \hfill (2.31)

Here $\lambda_1, \lambda_2, \ldots\ldots, \lambda_m$ are Lagrange Multipliers.

For the stationary points

$$\frac{\partial L}{\partial x_j} = 0 \,, \frac{\partial L}{\partial \lambda_j} = 0 \quad , \quad \text{for } j = 1, 2, \ldots., n; \ i = 1, 2, \ldots\ldots, m. \tag{2.32}$$

Chapter 3

A Fuzzy EOQ Model with Cost of Interest, Time Dependent Holding Cost, Without Shortages under a Space Constraint: A Fuzzy Geometric Programming and Non-Linear Programming Approach

Economic order quantity (EOQ) model plays important roles in the field of decision theory. **The basic objective of an inventory model is to minimize the total average cost or maximize the total average profit, also ensuring that the customer-seller relation dose not suffers at the same time.** *This model was introduced by F. D. Harris (1913). Later Hadley and Whitin (1963) analyzed various types of inventory models. In the past researchers assumed the parameters involved in an inventory model such as the demand, holding cost, se-up cost, deteriorating cost, etc. as a crisp values. But in reality some uncertainty occurs, i.e., in real life situation fuzzy set theory is more realistic than crisp set theory.*

In this chapter, an economic order quantity (EOQ) model with cost of interest, time dependent holding cost, without shortages is formulated and solved. In most real world situation, the objective and constraint function of the decision makers are imprecise in nature. Hence the coefficients are imposed here in fuzzy environment. Here the model is solved by geometric programming (GP), non-linear programming (NLP), fuzzy geometric programming (FGP), and fuzzy non linear programming (FNLP) technique respectively. Finally the model is illustrated by numerical examples and applications and we have seen that the fuzzy geometric-programming (FGP) technique gives better result than any other non-linear techniques.

3.1   Mathematical Model

An inventory model is developed under the following notations and assumptions.

3.1.1   Notations

I(t):Inventory level at any time, t≥0.

D: Constant demand per unit time.

T: Cycle of length.

S: Setup cost per batch.

H(t): Time dependent holding cost per unit quantity per unit time.

q: Production quantity per batch.

f(S,q): Total cost of interest per cycle.

TAC(D,S,q):Total average cost per unit time.

w_0: Space area per unit quantity.

W: Total storage space area.

3.1.2 Assumptions

a) The inventory system involves only one item.

b) The replenishment occurs instantaneously at infinite rate.

c) The lead time is negligible.

d) Demand rate is constant.

e) Total cost of interest is inversely related to set-up cost, production quantity and

directly related to cycle length, as $f(S,q) = bTS^{-x}q^{-y}$,   b, x, y$\in$ $\mathbb{R}$ (>0).

f) Holding cost is time dependent, $H(t) = H.at$,   a, H$\in$ $\mathbb{R}$ (>0).

3.1.3 Crisp inventory model

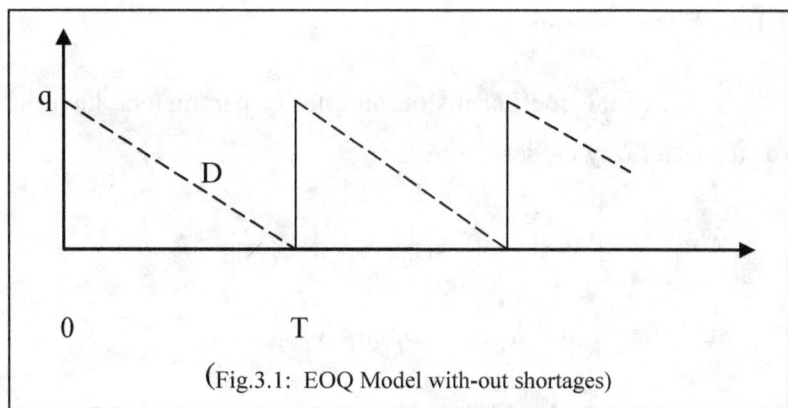

(Fig.3.1: EOQ Model with-out shortages)

The differential equations describing the above model as follows

$$\frac{dI(t)}{dt} = -D, \qquad 0 \le t \le T.$$ (3.1.3.1)

With the boundary condition $I(0) = q$, $I(T) = 0$.

The solution of (3.1.3.1) is obtained as

$$I(t) = q - Dt.$$ (3.1.3.2)

Also there are

$$T = q/D.$$

Now the inventory holding cost $= H\int_0^T at. I(t)dt = \frac{aHq^3}{6D^2}$ (3.1.3.3)

Total inventory related cost per cycle = set-up cost + holding cost + interest cost

$$= S + \frac{aHq^3}{6D^2} + f(S, q)$$ (3.1.3.4)

So total average cost per cycle is given by

$$TAC(D,S,q) = \frac{SD}{q} + \frac{aHq^2}{6D} + \frac{b}{S^x q^y}$$ (3.1.3.5)

And storage area $= w_0 q$.

Hench the inventory model can be written as

$$\text{Min} \qquad TAC(D,S,q) = \frac{SD}{q} + \frac{aHq^2}{6D} + \frac{b}{S^x q^y}$$ (3.1.3.6)

Subject to $w_0 q \leq W$, $\qquad D, S, q > 0$.

3.1.4 Fuzzy model

When the coefficients become fuzzy parameters, then the crisp model (3.1.3.6) can be written as a fuzzy model

$$\widetilde{Min} \qquad TAC(D,S,q) = \frac{SD}{q} + \frac{\tilde{a}\tilde{H}q^2}{6D} + \frac{\tilde{b}}{S^x q^y}$$

Subject to $\widetilde{w_0} q \lesssim W$, $\qquad D,S,q > 0$. (3.1.4.1)

3.2 Solution in Crisp Environment

3.2.1 Geometric programming method

Here the primal problem is

Min $\qquad$ TAC(D,S,q) $= \dfrac{SD}{q} + \dfrac{aHq^2}{6D} + \dfrac{b}{S^x q^y}$ $\qquad\qquad$ (3.2.1.1)

Subject to $w_0 q \leq W,$ $\qquad$ D, S, q > 0.

Corresponding dual problem of (3.2.1.1) is given by

Max d(ω) $= (\dfrac{1}{\omega_1})^{\omega_1}(\dfrac{aH}{6\omega_2})^{\omega_2}(\dfrac{b}{\omega_3})^{\omega_3}(\dfrac{w}{W\omega_4})^{\omega_4}\omega_{01}{}^{\omega_{01}}$ $\qquad\qquad$ (3.2.1.2)

Subject to

$\omega_1 + \omega_2 + \omega_3 = 1$

$\omega_1 - x\omega_3 = 0$

$\omega_1 - \omega_2 = 0$

$-\omega_1 + 2\omega_2 - y\omega_3 + \omega_{01} = 0$

$\omega_1, \omega_2, \omega_3, \omega_{01} \geq 0.$

Solving the above system of linear equations, we get

$$\omega_1 = \dfrac{x}{1+2x}, \omega_2 = \dfrac{x}{1+2x}, \omega_3 = \dfrac{1}{1+2x}, \omega_{01} = \dfrac{y-x}{1+2x}.$$

Putting these values to the objective function of the problem (3.2.1.2), we get

$$d(\omega) = (\dfrac{1+2x}{x})^{\frac{x}{1+2x}}(\dfrac{aH(1+2x)}{6x})^{\frac{x}{1+2x}}(\dfrac{b(1+2x)}{x})^{\frac{1}{1+2x}}(\dfrac{w_0(1+2x)}{W(y-x)})^{\frac{y-x}{1+2x}}(\dfrac{y-x}{1+2x})^{\frac{y-x}{1+2x}}$$

$$= (\dfrac{1+2x}{x}\dfrac{aH(1+2x)}{6x})^{\frac{x}{1+2x}}\left(\dfrac{b(1+2x)}{x}\right)^{\frac{1}{1+2x}}(\dfrac{w_0(1+2x)}{W(y-x)}\dfrac{y-x}{1+2x})^{\frac{y-x}{1+2x}}$$

From the relation between primal-dual variables, we get

$$\dfrac{SD}{q} = \dfrac{x}{1+2x}(\dfrac{1+2x}{x}\dfrac{aH(1+2x)}{6x})^{\frac{x}{1+2x}}\left(\dfrac{b(1+2x)}{x}\right)^{\frac{1}{1+2x}}(\dfrac{w_0(1+2x)}{W(y-x)}\dfrac{y-x}{1+2x})^{\frac{y-x}{1+2x}}) = \dfrac{x}{1+2x}k,$$

$$\dfrac{aHq^2}{6D} = \dfrac{x}{1+2x}(\dfrac{1+2x}{x}\dfrac{aH(1+2x)}{6x})^{\frac{x}{1+2x}}\left(\dfrac{b(1+2x)}{x}\right)^{\frac{1}{1+2x}}(\dfrac{w_0(1+2x)}{W(y-x)}\dfrac{y-x}{1+2x})^{\frac{y-x}{1+2x}}) = \dfrac{x}{1+2x}k,$$

$$\dfrac{b}{S^x q^y} = \dfrac{1}{1+2x}(\dfrac{1+2x}{x}\dfrac{aH(1+2x)}{6x})^{\frac{x}{1+2x}}\left(\dfrac{b(1+2x)}{x}\right)^{\frac{1}{1+2x}}(\dfrac{w_0(1+2x)}{W(y-x)}\dfrac{y-x}{1+2x})^{\frac{y-x}{1+2x}}) = \dfrac{1}{1+2x}k,$$

$$\frac{w_0 q}{W} = 1, \quad \text{where } k = \left(\frac{1+2x}{x} \frac{aH(1+2x)}{6x} \right)^{\frac{x}{1+2x}} \left(\frac{b(1+2x)}{x} \right)^{\frac{1}{1+2x}} \left(\frac{w_0(1+2x)}{W(y-x)} \frac{y-x}{1+2x} \right)^{\frac{y-x}{1+2x}}.$$

Solving above, we get

$$S = \frac{6 w_0 k^2}{aHW} \left(\frac{x}{1+2x} \right)^2,$$

$$D = \frac{aH W^2 k^{-1}}{6 w_0{}^2} \left(\frac{1+2x}{x} \right),$$

$$q = \frac{W}{w_0}. \tag{3.2.1.3}$$

3.2.2 Non-linear programming method

The problem (3.1.3.6) can be written as Lagrangian form

$$\mathcal{L}(S, D, q) = \frac{SD}{q} + \frac{aH q^2}{6D} + \frac{b}{S^x q^y} - \lambda(W - q w_0) \tag{3.2.2.1}$$

From Kuhn-Tucker necessary conditions the solution can be obtained as (doing the partially derivative w.r.t. S, D, q and λ respectively),

$$\frac{D}{q} - \frac{xb}{S^{x+1} q^2} = 0,$$

$$\frac{S}{q} - \frac{aH q^2}{6D^2} = 0,$$

$$-\frac{SD}{q^2} + \frac{aH q}{3D} - \frac{yb}{S q^{y+1}} + \lambda w_0 = 0,$$

and $\lambda(W - w_0 q) = 0.$ \hfill (3.2.2.2)

Here two conditions $\lambda = 0$ or $W - w_0 q = 0$.

When $\lambda = 0$, does not satisfies all the equation of (3.2.2.2), so $W - w_0 q = 0$.

And optimal solution is,

$$S^* = \frac{aH(xb)^{\frac{2}{2x+1}}}{6 \frac{W}{w_0}^{\frac{5}{2x+1}} \left(\frac{aH}{6} \right)^{\frac{2x+2}{2x+1}}},$$

$$D^* = \frac{\left(\frac{W}{w_0} \right)^{\frac{3x+4}{2x+1}} \left(\frac{aH}{6} \right)^{\frac{x+1}{2x+1}}}{(xb)^{\frac{1}{2x+1}}}, \tag{3.2.2.3}$$

$$q^* = \frac{W}{w_0} \ .$$

3.3   Solution Procedure in Fuzzy Environment

3.3.1 Mathematical analysis:

Consider a non-linear programming as follows,

Min $g_0(x)$

Subject to $g_i(x) \leq 1$ $(1 \leq i \leq n)$,   (3.3.1.1)

              $x > 0$.

Its objective and constraints of the form

$g_i(x) = \sum_{k=1}^{T_i} C_{ik} \prod_{j=1}^{m} x_j^{\alpha_{ikj}}$ $(0 \leq i \leq n)$

              $x_j > 0$, $(j = 1, 2, \ldots, m)$.

Here $c_{ik} (> 0)$, $(k = 1, 2, \ldots, T_0)$ and α_{ikj} be any real numbers.

When the objective and constraint goals, coefficients and exponents become fuzzy sets and fuzzy numbers respectively, then we transform (P) into a fuzzy geometric programming as follows,

$\widetilde{Min}$ $g_0(x)$

Subject to $g_i(x) \lesssim 1$ $(1 \leq i \leq n)$ (3.3.1.2)

            $x > 0$,

Its objective and constraints of the form $g_i(x) = \sum_{k=1}^{T_i} \tilde{c}_{ik} \prod_{j=1}^{m} x_j^{\tilde{\alpha}_{ikj}}$ $(0 \leq i \leq n)$, are all posynomials of x in which coefficients $\tilde{c}_{ik}$ and indices $\tilde{\alpha}_{ikj}$ are fuzzy numbers.

3.3.2 Some definitions and theorems

Definition 3.3.2.1 For nth parabolic flat fuzzy number $(a_1, a_2, a_3, a_4)_{PfFN}$ containing the coefficients $\tilde{c}_{ik}$ $(0 \leq i \leq n; 1 \leq k \leq T_i)$, then the membership function of $\tilde{c}_{ik}$ is

$$\mu_{\tilde{c}_{ik}}(\tilde{c}_{ik}) = \begin{cases} 1 - (\frac{a_2 - c_{ik}}{a_2 - a_1})^n & for\, a_1 \leq c_{ik} \leq a_2 \\ 1 & for\, a_1 \leq c_{ik} \leq a_2 \\ 1 - (\frac{c_{ik} - a_3}{a_4 - a_3})^n & for\, a_3 \leq c_{ik} \leq a_4 \\ 0 & for\, otherwise. \end{cases}$$
(3.3.2.1.1)

Similarly, we can determine the membership function of the indexes $\tilde{\alpha}_{ikj}$ $(0 \leq i \leq n; 1 \leq k \leq T_i; 1 \leq j \leq m)$.

Note:

a)   when n=1, $\tilde{c}_{ik}$ become Trapezodial Fuzzy Number (TrFN),

b)   when n=1, and $a_3 = a_4$, $\tilde{c}_{ik}$ become Triangular Fuzzy Number (TFN),

c)   when n=2, $\tilde{c}_{ik}$ become Parabolic flat Fuzzy Number (PfFN),

d)   when n=2, and $a_3 = a_4$, $\tilde{c}_{ik}$ become Parabolic Fuzzy Number (PrFN).

Definition 3.3.2.2 Here δ-cut of $\tilde{c}_{ik}$ $(0 \leq i \leq n; 1 \leq k \leq T_i)$ is given by

$$\mu_{\tilde{c}_{ik}}^{-1}(\delta) = [\mu_{\tilde{c}_{ikL}}^{-1}(\delta), \mu_{\tilde{c}_{ikR}}^{-1}(\delta)] = [a_1 + \sqrt[n]{1-\delta}(a_2 - a_1), a_4 - \sqrt[n]{1-\delta}(a_4 - a_3)].$$
(3.3.2.2.1)

Similarly, we can determine the δ-cut of $\tilde{\alpha}_{ikj}$ $(0 \leq i \leq n; 1 \leq k \leq T_i; 1 \leq j \leq m)$.

Proposition 3.3.2.3 When the coefficient and indexes of the fuzzy geometric programming problem are taken as fuzzy numbers

$\widetilde{Min} \qquad \sum_{k=1}^{T_i} \tilde{c}_{ok} \prod_{j=1}^{m} x_j^{\tilde{\alpha}_{okj}}$

Subject to $\sum_{k=1}^{T_i} \tilde{c}_{ik} \prod_{j=1}^{m} x_j^{\tilde{\alpha}_{ikj}} \lesssim 1$ $\qquad\qquad (1 \leq i \leq n)$, $\qquad\qquad$ (3.3.2.3.1)

$\qquad x_j > 0.$

Using δ-cut of fuzzy numbers coefficients and indexes, the above problem is reduces to

$\widetilde{Min} \sum_{k=1}^{T_0} [\mu_{\tilde{c}_{okL}}^{-1}(\delta), \mu_{\tilde{c}_{okR}}^{-1}(\delta)] \prod_{j=1}^{m} x_j^{[\mu_{\tilde{\alpha}_{okjL}}^{-1}(\delta), \mu_{\tilde{\alpha}_{okjR}}^{-1}(\delta)]}$

Subject to $\sum_{k=1}^{T_i} [\mu_{\tilde{c}_{ikL}}^{-1}(\delta), \mu_{\tilde{c}_{ikR}}^{-1}(\delta)] \prod_{j=1}^{m} x_j^{[\mu_{\tilde{\alpha}_{ikjL}}^{-1}(\delta), \mu_{\tilde{\alpha}_{ikjR}}^{-1}(\delta)]} \lesssim 1$ $\quad (1 \leq i \leq n)$,

$\qquad x_j > 0.$

Which is equivalent to

$\widetilde{Min} \qquad \sum_{k=1}^{T_o} \mu_{\tilde{c}_{okL}}^{-1}(\delta) \prod_{j=1}^{m} x_j^{\mu_{\tilde{\alpha}_{okjS}}^{-1}(\delta)}$

Subject to $\sum_{k=1}^{T_i} \mu_{\tilde{c}_{ikL}}{}^{-1}(\delta) \prod_{j=1}^{m} x_j{}^{\mu_{\tilde{\alpha}_{ikjS}}{}^{-1}}(\delta) \leq 1,$ $\quad (1\leq i\leq n).$ $\qquad (3.3.2.3.2)$

Where

$$\mu_{\tilde{\alpha}_{ikjS}}{}^{-1}(\delta) = \begin{cases} \mu_{\tilde{\alpha}_{ikjL}}{}^{-1}(\delta) & when\,\tilde{\alpha}_{ikjL} > 0, \\ \\ \mu_{\tilde{\alpha}_{ikjR}}{}^{-1}(\delta) & when\,\tilde{\alpha}_{ikjL} < 0. \end{cases} \qquad (1\leq i\leq n)$$

Definition 3.3.2.4 For any $x \in \mathbb{R}^m$ and feasible index $d_i \in \mathbb{R}$ ($\mathbb{R}$ is the real number set), if $g_i(x,\delta) = \sum_{k=1}^{T_i} \mu_{\tilde{c}_{ikL}}{}^{-1}(\delta) \prod_{j=1}^{m} x_j{}^{\mu_{\tilde{\alpha}_{ikjS}}{}^{-1}}(\delta) \leq 1 (1\leq i \leq n)$, then the linear membership function are given by

$$\mu_0(g_0(x,\delta)) = \begin{cases} 1 & if\,g_0(x,\delta) \leq z_0, \\ \left(\frac{z_0+d_0-g_0(x,\delta)}{d_0}\right) & if\,z_0 \leq g_0(x,\delta) \leq z_0 + d_0, \\ 0 & if\,g_0(x,\delta) \geq z_0 + d_0, \end{cases} \qquad (3.3.2.4.1)$$

$$\mu_i(g_i(x,\delta)) = \begin{cases} 1 & if\,g_i(x,\delta) \leq z_0, \\ \left(\frac{1+d_i-i(x,\delta)}{d_0}\right) & if\,1 \leq g_i(x,\delta) \leq 1 + d_i, \\ 0 & if\,g_i(x,\delta) \geq 1 + d_i, \end{cases} \qquad (3.3.2.4.2)$$

Based on Zimmerman, first finding δ-cut of the fuzzy numbers in coefficients and indexes then we built   membership functions of both objective and constraints goals and using max-min operator the above problem (3.3.2.3.2) reduced to a fuzzy non-linear programming(FNLP) problem

Max $\qquad \lambda$

Subject to $\quad \mu_i(\sum_{k=1}^{T_i} \mu_{\tilde{c}_{ikL}}{}^{-1}(\delta) \prod_{j=1}^{m} x_j{}^{\mu_{\tilde{\alpha}_{ikjS}}{}^{-1}}(\delta)) \geq \lambda(1\leq i\leq n),$ $\qquad (3.3.2.4.3)$

$\qquad x > 0, \quad \lambda, \delta \in [0,1].$

Which is equivalent to a geometric programming (GP) problem with parameters λ, δ.

Min $\qquad \lambda^{-1}$

Subject to $\quad \mu_i(\sum_{k=1}^{T_i} \mu_{\tilde{c}_{ikL}}{}^{-1}(\delta) \prod_{j=1}^{m} x_j{}^{\mu_{\tilde{\alpha}_{ikjS}}{}^{-1}}(\delta)) \geq \lambda(1\leq i\leq n),$ $\qquad (3.3.2.4.4)$

$\qquad x > 0, \quad \lambda, \delta \in [0,1],$

Theorem 3.3.2.5

Let the membership function $\mu_i(g_i(x,\delta))$, $\mu_{\tilde{c}_{ik}}(c_{ik})$, $\mu_{\tilde{\alpha}_{ikj}}(\alpha_{ikj})$ be all continuous and strictly monotone. Then (3.3.2.4.4) is equivalent with

Min $\qquad \lambda^{-1}$

Subject to $\qquad \dfrac{\sum_{k=1}^{T_i} \mu_{\tilde{c}_{ikL}}{}^{-1}(\delta) \prod_{j=1}^{m} x_j{}^{\mu_{\tilde{\alpha}_{ikjS}}{}^{-1}(\delta)}}{\mu_i{}^{-1}(\delta)} \leq 1,$ $\qquad\qquad$ (3.3.2.5.1)

$\qquad\qquad x > 0, \quad \lambda, \delta \in [0,1], \quad (0 \leq i \leq n, \ 1 \leq j \leq m).$

Proof: Pls. see the reference S. Islam, T.K. Roy (2006).

Corollary 3.3.2.6: Let the membership function $\mu_i(g_i(x,\delta))$, $\mu_{\tilde{c}_{ik}}(c_{ik})$, $\mu_{\tilde{\alpha}_{ikj}}(\alpha_{ikj})$ be all continuous and strictly monotone and the problem is

Min $\qquad \lambda^{-1}$

Subject to $\dfrac{\sum_{k=1}^{T_i} \mu_{\tilde{c}_{ikL}}{}^{-1}(\delta) \prod_{j=1}^{m} x_j{}^{\mu_{\tilde{\alpha}_{ikjS}}{}^{-1}(\delta)}}{\mu_i{}^{-1}(\delta)} \leq 1,$ $\qquad\qquad$ (3.3.2.6.1)

$\qquad\qquad x > 0, \quad \lambda, \delta \in [0,1], \quad (0 \leq i \leq n, \ 1 \leq j \leq m).$

Which is a classical posynomial geometric programming with parameters γ, δ.

Its dual form is

Max $\qquad d(\omega) = \left(\dfrac{\lambda^{-1}}{\omega_{00}}\right)^{\omega_{00}} \prod_{i=0}^{n} \prod_{k=1}^{T_i} \left(\dfrac{\mu_{\tilde{c}_{ik}}{}^{-1}(\delta)/\mu_i{}^{-1}(\lambda)}{\omega_{ik}}\right)^{\omega_{ik}}$

Subject to $\omega_{00} = 1,$

$\qquad\qquad \omega_{00} = \sum_{k=1}^{T_0} \omega_{0k},$

$\qquad\qquad (\Gamma(\delta))^T \omega = 0, \qquad \lambda, \delta \in [0, 1],$ $\qquad\qquad$ (3.3.2.6.2)

$\qquad\qquad \omega \geq 0.$

Where $\omega_{ik} = \omega_{ik}(\delta, \lambda),$

and $\Gamma(\delta) = \begin{pmatrix} \tilde{\alpha}_{011}{}^{-1}(\delta) \ldots & \tilde{\alpha}_{01l}{}^{-1}(\delta) \ldots & \tilde{\alpha}_{01m}{}^{-1}(\delta) \\ \ldots & \ldots & \ldots \\ \tilde{\alpha}_{0J_01}{}^{-1}(\delta) \ldots & \tilde{\alpha}_{0J_01}{}^{-1}(\delta) \ldots & \tilde{\alpha}_{0J_01}{}^{-1}(\delta) \\ \ldots \ldots \ldots & & \\ \tilde{\alpha}_{p11}{}^{-1}(\delta) \ldots & \tilde{\alpha}_{p1l}{}^{-1}(\delta) \ldots & \tilde{\alpha}_{p1m}{}^{-1}(\delta) \\ \ldots & \ldots & \ldots \\ \tilde{\alpha}_{pJ_p1}{}^{-1}(\delta) \ldots & \tilde{\alpha}_{pJ_p1}{}^{-1}(\delta) \ldots & \tilde{\alpha}_{pJ_p1}{}^{-1}(\delta) \end{pmatrix}.$

3.4   Solution procedure by FGP method

When coefficient and exponents are taken as a triangular fuzzy number i.e., in general $\tilde{\mathcal{L}} = (a_1, a_2, a_3)$. Then the δ-cut of the fuzzy number $\tilde{\mathcal{L}}$, is given by $\mathcal{L}(\delta) = [a_1 + \delta(a_2 - a_1), a_3 - \delta(a_3 - a_2)]$, $\delta \in [0,1]$.

Taking the membership function as in (3.3.2.4.1) and (3.3.2.4.2), turn the problem (3.1.4.1) into (3.3.2.6.1) and obtain

Min λ^{-1}

Subject to $\dfrac{-SDq^{-1} - \left(H^1 + \delta(H^2 - H^1)\right)aq^2 D^{-1}/6 - (b^1 + \delta(b^2 - b^1))S^{-x}Q^{-y}}{(-(z_0 + d_0 - 1) + d_0\lambda)} \leq 1$ (3.4.1)

$\dfrac{(w^1 + \delta(w^2 - w^1))q}{W + d_1\lambda} \leq 1,$

$D, S, q > 0, \qquad \gamma, \delta \in [0, 1].$

The dual form of (3.4.1) is given by

Max $d(\omega) = \left(\dfrac{\lambda^{-1}}{\omega_{00}}\right)^{\omega_{00}} \left(\dfrac{\frac{1}{\omega_{01}}}{(z_0 + d_0 - 1) - d_0\lambda}\right)^{\omega_{01}} \left(\dfrac{\frac{(H^1 + \delta(H^2 - H^1))a}{6\omega_{02}}}{(z_0 + d_0 - 1) - d_0\lambda}\right)^{\omega_{02}} \left(\dfrac{\frac{(b^1 + \delta(b^2 - b^1))}{\omega_{03}}}{(z_0 + d_0 - 1) - d_0\lambda}\right)^{\omega_{03}} \left(\dfrac{\frac{(w_0^1 + \delta(w_0^2 - w_0^1))}{\omega_{11}}}{W + d_1\lambda}\right)^{\omega_{11}} (w_{11})^{w_{11}}$

Subject to

$\omega_{00} = 1,$

$\omega_{01} + \omega_{02} + \omega_{03} + \omega_{11} = \omega_{00},$ (3.4.2)

$\omega_{01} - x\omega_{03} = 0,$

$\omega_{01} - \omega_{02} = 0,$

$-\omega_{01} + 2\omega_{02} - y\omega_{03} + \omega_{11} = 0,$

$\omega_{01}, \omega_{02}, \omega_{03}, \omega_{11} \geq 0.$

From (3.4.2) we get $\omega_{01} = \dfrac{x}{1+2x}, \omega_{02} = \dfrac{x}{1+2x}, \omega_{03} = \dfrac{1}{1+2x}, \omega_{11} = \dfrac{y-x}{1+2x}.$

Putting the values of the objective functions of the problem (3.4.2), we get

$$d(\omega) = \lambda^{-1} \left(\frac{\frac{1+2x}{x}}{(z_0+d_0-1)-d_0\lambda} \right)^{\frac{x}{1+2x}} \left(\frac{\frac{\left(H^1+\delta(H^2-H^1)\right)a(1+2x)}{6x}}{(z_0+d_0-1)-d_0\lambda} \right)^{\frac{x}{1+2x}} \left(\frac{\left(b^1+\delta(b^2-b^1)(1+2x)\right)}{(z_0+d_0-1)-d_0\lambda} \right)^{\frac{1}{1+2x}}$$

$$\times \left(\frac{3(w_0{}^1+\delta(w_0{}^2-w_0{}^1)(1+2x)/(y-x)}{W+d_1\lambda} \right)^{\frac{y-x}{1+2x}} \left(\frac{y-x}{1+2x} \right)^{\frac{y-x}{1+2x}}.$$

We can obtained λ by the aid of $d(\omega) = \lambda^{-1}$. Then the above equation is reduces to

$$\left(\frac{\frac{1+2x}{x}}{(z_0+d_0-1)-d_0\lambda} \right)^{\frac{x}{1+2x}} \left(\frac{\frac{\left(H^1+\delta(H^2-H^1)\right)a(1+2x)}{6x}}{(z_0+d_0-1)-d_0\lambda} \right)^{\frac{x}{1+2x}} \left(\frac{(b^1+\delta(b^2-b^1)(1+2x)}{(z_0+d_0-1)-d_0\lambda}\right)^{\frac{1}{1+2x}} \left(\frac{3\left(\frac{w_0{}^1+\delta(w_0{}^2-w_0{}^1)(1+2x)}{(y-x)}\right)}{W+d_1\lambda} \right)^{\frac{y-x}{1+2x}}$$

$$\times \left(\frac{y-x}{1+2x} \right)^{\frac{y-x}{1+2x}} = 1 \tag{3.4.3}$$

Solving the above non-linear equation of γ for given $\delta \in [0, 1]$ by Newton-Raphson method, we obtain the value of λ^*. Putting the value of λ^*, we obtained the value of the dual objective function.

Again from the relation between primal-dual variables, we get

$$\frac{-SDq^{-1}}{(-(z_0+d_0-1)+d_0\lambda^*)} = \frac{\omega_{01}{}^*}{\omega_{00}{}^*} = \frac{x}{1+2x},$$

$$\frac{-\left(H^1+\delta(H^2-H^1)\right)aq^2D^{-1}/6}{(-(z_0+d_0-1)+d_0\lambda^*)} = \frac{\omega_{02}{}^*}{\omega_{00}{}^*} = \frac{x}{1+2x},$$

$$\frac{-(b^1+\delta(b^2-b^1))S^{-1}q^{-2}}{(-(z_0+d_0-1)+d_0\lambda^*)} = \frac{\omega_{03}{}^*}{\omega_{00}{}^*} = \frac{1}{1+2x},$$

$$\frac{(w_0{}^1+\delta(w_0{}^2-w_0{}^1))q}{W+d_1\lambda^*} = \frac{\omega_{11}{}^*}{\omega_{11}{}^*}. \tag{3.4.4}$$

3.5   Solution Procedure by FNLP method

A fuzzy non-linear programming problem with fuzzy resources and objective are defined as

$$\widetilde{Min} \qquad g_0(x)$$

$$\text{Subject to } g_i(x) \lesssim 1 \quad (1 \leq i \leq n) \tag{3.5.1}$$

$$x > 0,$$

Its objective and constraints, $g_i(x) = \sum_{k=1}^{T_i} \tilde{c}_{ik} \prod_{j=1}^{m} x_j{}^{\tilde{\alpha}_{ikj}}$ $(0 \leq i \leq n)$, are all posynomials of x in which coefficients $\tilde{c}_{ik}$ and index $\tilde{\alpha}_{ikj}$ are fuzzy numbers. In fuzzy set theory, the fuzzy objective and fuzzy resources are obtained by their membership functions, which may be linear or nonlinear. Here $\mu_0(g_0(x,\delta))$ and $\mu_i(g_i(x,\delta))$ (i = 1, 2 ...., m) are assumed to be non-increasing continuous linear membership

$$\mu_0(g_0(x,\delta)) = \begin{cases} 1 & if\, g_0(x,\delta) \le z_0 \,, \\ \left(\frac{z_0+d_0-g_0(x,\delta)}{d_0}\right) & if\, z_0 \le g_0(x,\delta) \le z_0 + d_0, \\ 0 & if\, g_0(x,\delta) \ge z_0 + d_0, \end{cases} \tag{3.5.2}$$

$$\mu_i(g_i(x,\delta)) = \begin{cases} 1 & if\, g_i(x,\delta) \le z_0, \\ \left(\frac{1+d_i-i(x,\delta)}{d_i}\right) & if\; 1 \le g_i(x,\delta) \le 1 + d_i, \\ 0 & if\, g_i(x,\delta) \ge 1 + d_i, \end{cases} \tag{3.5.3}$$

In this formulation, the fuzzy objective goal tolerance is d_0 and corresponding tolerance is d_i ((i = 1, 2 ...., m). To solve the problem (3.1.4.1), the max-min operator of Bellman and approach of Zimmerman are implemented. Taking the membership function as in (3.5.2) and (3.5.3), turn the problem (3.1.4.1) into (3.3.2.6.1), and obtain

Maximum α

Such that $\mu_0(x) \ge \alpha$

$$\mu_i(x) \ge \alpha \qquad (i = 1, 2,\ldots\ldots, m) \tag{3.5.4}$$

$x \ge 0, \delta \in [0,1]$.

A new function, i.e., the Lagrangian function $\mathcal{L}(\alpha, x, \lambda)$ is formed by introducing (m+1) Lagrangian multipliers $\lambda = (\lambda_0, \lambda_1, \ldots\ldots, \lambda_m)$.

$\mathcal{L}(\alpha, x, \lambda) = \alpha - \sum_{i=0}^{m} \lambda_i(g_i(x) - b_i - (1-\alpha)d_i)$. The necessary condition of Khun-Tucker for the optimal solution to this given problem implies that the optimal values $x_1^*, x_2^*, \ldots\ldots, x_n^*$, and $\lambda_1^*, \lambda_2^*, \ldots\ldots, \lambda_m^*$ should satisfy

$$\frac{\partial \mathcal{L}}{\partial x_j} = 0 \qquad j = 1, 2, \ldots, n,$$

$$\frac{\partial \mathcal{L}}{\partial \alpha} = 0$$

$$\lambda_i(g_i(x) - b_i - (1-\alpha)d_i = 0$$

$$g_i(x) \le b_i + (1-\alpha)d_i.$$

So the problem (3.2.3.7) can be transformed as Lagrangian form as

$$\mathcal{L}(\alpha, S, D, q, \lambda_1, \lambda_2) = \alpha - \lambda_1 \left(\frac{SD}{q} + \frac{aHq^2}{6D} + \frac{b}{S^x q^y} - z_0 - (1-\alpha)d_0 \right) - \lambda_2 (qw_0 - W - (1-\alpha)d_1) \quad (3.5.5)$$

Where objective goal is z_0 with tolerance d_0 and space constraint goal is W with tolerance d_1.

From the Khun-Tucker necessary conditions

$$\frac{\partial \mathcal{L}}{\partial \alpha} = 1 - \lambda_1 d_0 - \lambda_2 d_1 \geq 0, \qquad\qquad \text{and} \quad \alpha(1 - \lambda_1 d_0 - \lambda_2 d_1) = 0,$$

$$\frac{\partial \mathcal{L}}{\partial S} = -\lambda_1 \left(\frac{D}{q} - \frac{xb}{S^{x+1}q^2} \right) \leq 0, \qquad\qquad S\left(\frac{D}{q} - \frac{xb}{S^{x+1}q^2} \right) = 0,$$

$$\frac{\partial}{\partial D}\mathcal{L} = -\lambda_1 \left(\frac{S}{q} - \frac{aHq^2}{6D^2} \right) \leq 0, \qquad\qquad D\left(\frac{S}{q} - \frac{aHq^2}{6D^2} \right) = 0,$$

$$\frac{\partial \mathcal{L}}{\partial q} = -\lambda_1 \left(-\frac{SD}{q^2} + \frac{aHq}{3D} - \frac{yb}{Sq^{y+1}} - \lambda_2 w_0 \right) \leq 0, \qquad\qquad q\lambda_1 \left(-\frac{SD}{q^2} + \frac{aHq}{3D} - \frac{yb}{Sq^{y+1}} - \lambda_2 w_0 \right) = 0,$$

$$\frac{\partial \mathcal{L}}{\partial \lambda_1} = \left(\frac{SD}{q} + \frac{aHq^2}{6D} + \frac{b}{S^x q^y} - z_0 - (1-\alpha)d_0 \right) \leq 0, \qquad \lambda_1 \left(\frac{SD}{q} + \frac{aHq^2}{6D} + \frac{b}{S^x q^y} - z_0 - (1-\alpha)d_0 \right) = 0,$$

$$\frac{\partial \mathcal{L}}{\partial \lambda_2} = (w_0 q - W - (1-\alpha)d_1) \leq 0, \qquad\qquad \lambda_2(w_0 q - W - (1-\alpha)d_1) = 0.$$

And solution is

$$S^* = \frac{aH(xb)^{\frac{2}{2x+1}}}{\frac{6(W+(1-\alpha^*)d_1)^{\frac{5}{2x+1}}}{w_0}(\frac{aH}{6})^{\frac{2x+2}{2x+1}}},$$

$$D^* = \frac{(\frac{W+(1-\alpha^*)d_1}{w_0})^{\frac{3x+4}{2x+1}}(\frac{aH}{6})^{\frac{x+1}{2x+1}}}{(xb)^{\frac{1}{2x+1}}}, \qquad\qquad (3.5.6)$$

$$q^* = \frac{W+(1-\alpha^*)d_1}{w_0}.$$

Where α^* is a root of $\left(\frac{S^*D}{q} + \frac{aHq^{2*}}{6D^*} + \frac{b}{S^{*x}q^{*y}} - z_0 - (1-\alpha)d_0 \right) = 0.$ \quad (3.5.7)

3.6 Numerical Example

Let the input value of the model (3.1.3.6) is

Table-3.1 (Crisp input data of the given model)

a	b	H	x	y	w_0	W
7	5	3	1	2	216	1000

Then the model is of the form

$$\text{Min} \qquad \text{TAC(D,S,q)} = \frac{SD}{q} + \frac{21q^2}{6D} + \frac{5}{Sq^2}$$

Subject to $216q \leq 1000,$

$D, S, q > 0.$ (3.6.1)

And corresponding solution in crisp environment, as follows

Table-3.2 (Optimal solution of the given for crisp model)

Crisp model	S^*	D^*	q^*	T^*	$TAC^*(S^*,D^*,q^*)$
GP	0.150	48.157	4.630	0.096	4.673
NLP	0.149	48.158	4.630	0.096	4.672

For fuzzy geometric programming method, lets we consider $z_0 = 4.63$, fuzzy objective goal tolerance $d_0 = 1$ and total storage space area tolerance $d_1 = 200$ H = (2,3,4), b = (4,5,6), w_0 = (214,216,218)(as a fuzzy triangular number) taking $\delta = 0.5$, then from (3.5.2) we get $\lambda = 0.087$.

For fuzzy non-linear programming method, lets we consider $z_0 = 4.63$, fuzzy objective goal $d_0 = 1$ and total storage space area tolerance $d_1 = 200$ from (7.6.4) we get $\alpha^* = 0.967$.

And corresponding solution in fuzzy environment, as follows

Table-3.3 (Optimal solution of the given fuzzy model)

Crisp model	S^*	D^*	q^*	T^*	$TAC^*(S^*,D^*,q^*)$
FGP	0.133	53.878	4.732	0.088	4.648
FNLP	0.148	48.304	4.660	0.096	4.663

From table-3.2 and table-3.3, we have seen that fuzzy geometric programming (FGP) is given better result than any others methods.

3.7   Sensitivity Analysis

3.7.1   We now examine to sensitivity analysis of the optimal solution of the model for changing a, b, H, and w_0 respectively, keeping the other parameters is unchanged. The initial are data from the above numerical example.

Table-3.4 (Sensitivity analysis)

Parameter	% of change	S^*	D^*	q^*	$TAC^*(S^*, D^*, q^*)$
a=3.50	-50	0.189	30.337	4.630	4.946
a=5.25	-25	0.165	39.753	4.630	4.718
a=7.00	0	0.150	53.878	4.630	4.673
a=8.75	25	0.139	55.881	4.630	4.698
a=10.5	50	0.131	63.104	4.630	4.755
b=2.50	-50	0.094	60.675	4.630	4.950
b=3.75	-25	0.124	53.004	4.630	4.716
b=5.00	0	0.150	53.878	4.630	4.673
b=6.25	25	0.173	44.705	4.630	4.697
b=7.50	50	0.196	42.069	4.630	4.754
H=1.50	-50	0.189	30.337	4.630	4.946
H=2.25	-25	0.165	39.753	4.630	4.718
H=3.00	0	0.150	53.878	4.630	4.673
H=3.75	25	0.139	55.882	4.630	4.698
H=4.50	50	0.131	63.104	4.630	4.755
w_0=108	-50	0.047	242.698	9.259	3.709
w_0=162	-25	0.093	94.229	6.173	4.246
w_0=216	0	0.150	53.878	4.630	4.673
w_0=270	25	0.217	28.611	3.704	5.034
w_0=324	50	0.294	18.697	3.086	5.350

Here we have given a rough graph, which shown how change the value of $TAC^*(S^*,$ $D^*, q^*)$ for different values of a, b, H andw_0 respectively by G.P method solution.

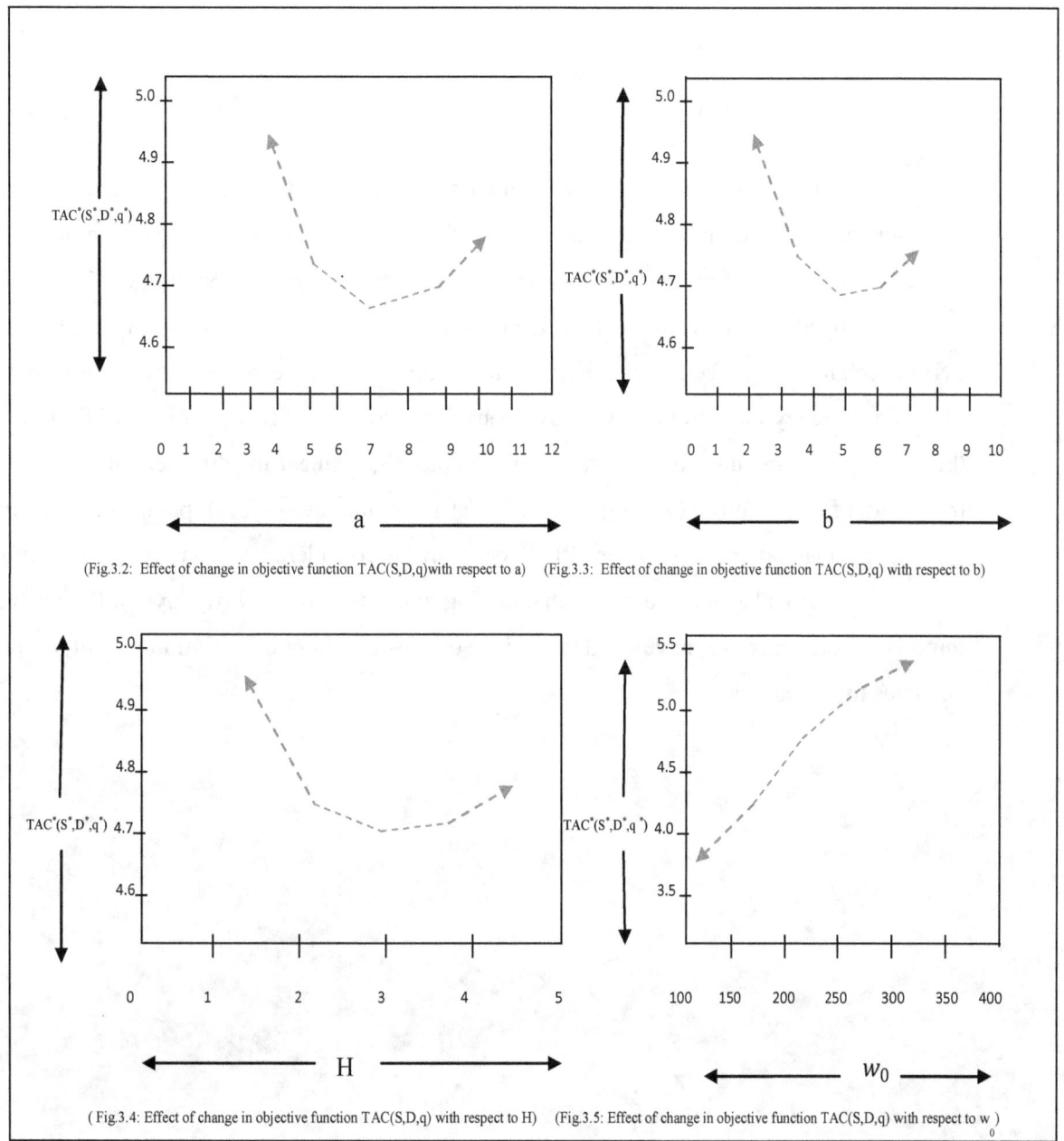

(Fig.3.2: Effect of change in objective function TAC(S,D,q)with respect to a)   (Fig.3.3: Effect of change in objective function TAC(S,D,q) with respect to b)

(Fig.3.4: Effect of change in objective function TAC(S,D,q) with respect to H)   (Fig.3.5: Effect of change in objective function TAC(S,D,q) with respect to w)

3.7.2 Effect, for increment of parameters-:

i)   For increasing value of "a", S decrease, D increase, q un-change, TAC(S, D, q) firstly decrease and then increase.

ii)   For increasing value of "b", S increase, D decrease, q un-change, TAC(S, D, q) firstly decrease and then increase.

iii)   For increasing value of "H", S decrease, D increase, q un-change, TAC(S, D, q) firstly decrease and then increase.

iv)   For increasing value of "w_0", S increase, D decrease, q decrease, TAC(S, D, q) increase.

3.8   Conclusion

In this chapter, we have presented a real life inventory problem in a crisp and fuzzy environment and presented solution along with sensitivity analysis. The inventory model is developed with cost of interest, time dependent holding cost, without shortages. This model has been developed for single item. In this chapter, the model is solved by GP, NLP, FGP and FNLP techniques respectively. Also we have given a real example and solved it various methods. In fuzzy environment, we have considered triangular fuzzy number (TFN). In future, the other type of membership functions such as piecewise linear hyperbolic, L-R fuzzy number, trapezoidal fuzzy number (TrFN), parabolic flat fuzzy number (PfFN), parabolic fuzzy number (PrFN), pentagonal fuzzy number (PFN) etc. can be considered to construct the membership function and model would be more challenging and interesting. Nowadays, inflation plays an important role buts we have neglected it. So consideration of inflation in future problem becomes more realistic.

Chapter 4

Fuzzy EOQ Model for Deteriorating Items, With Constant Demand, Shortages, and Fully Backlogging

Deterioration plays an important role in an inventory model. Deterioration is the gradual loss or damage of certain items such that the item can-not be used in future. Drugs, foods, medicines etc. are deteriorations items. The losses of deteriorations items must be taken into account when analyzing the inventory system.

In this chapter we have analyzed a fuzzy inventory model, for deterioration item with constant demand. Shortages are allowed under fully backlogged. Fixed cost, deterioration cost, shortages cost, holding cost are considered in this model. Fuzziness is applying by allowing the cost components (holding cost, deterioration, shortages cost, etc.). In fuzzy environment, we have considered all required parameter to be triangular fuzzy numbers (TFN). Here we have used nearest interval approximation (NIA) method to convert a triangular fuzzy number (TFN) to an interval number and transformed this interval number to a parametric interval-valued programming form. Finally several numerical examples are given for this model.

*Here we have considered **a special condition** (flood, strike, earthquake, etc.), when **demand falls to zero** in a time interval ($t_0 \leq t \leq t_e$) and considered three cases-*

*1) Demand falls to zero, **before the deterioration set in.***

*2) Demand falls to zero, **after the deterioration set in.***

*3) **Without accounts of any special condition.***

*Three different models are developed under these three cases in the crisps sense. There after we have transformed theses crisp inventory models to fuzzy inventory models. Here we have given several numerical examples and compared the result, it seen that $Tac_3(t_1) \leq Tac_1(t_1) \leq Tac_2(t_1)$ and conclude that the total average cost function maximum when demand falls to zero **after deteriorations start,** it minimum when no such unexpected condition arise.*

4.1 Mathematical Model

An economic order quantity (EOQ) model is developed under the following notations and assumptions.

4.1.1 Notations

$I(t)$: Inventory level at any time, $t \geq 0$.

T: Cycle of length.

t_0: Time point, when demand falls to zero (unexpected conditions arise).

t_e: Time point, when demand again start.

t_d: Time point, when the deterioration set in.

t_1: Time point, when inventory level falls to zero.

c_f: Fixed cost.

c_s: Shortages cost per unit per unit time.

c_d: Deteriorating cost per unit per unit time.

c_h: Holding cost per unit per unit time.

Q: Stock level at the beginning of the cycle.

$Tac_i(t_1)$: Total average cost per unit, $i = 1, 2, 3$.

$\tilde{c}_f$: Fuzzy fixed cost.

$\tilde{c}_s$: Fuzzy shortage cost per unit per unit time.

$\tilde{c}_d$: Fuzzy deteriorating cost per unit pee unit time.

$\tilde{c}_h$: Fuzzy holding cost per unit per unit time.

$\widetilde{Tac_i}(t_1)$: Fuzzy total average cost per unit, $i = 1, 2, 3$.

4.1.2 Assumptions

a) The inventory system involves only one item.

b) The replenishment occurs instantaneously at infinite rate.

c) The lead time is negligible.

d) Demand rate is constant D.

e) $2\theta t$ is deterioration rate per unit time per cycle, "θ" is constant.

4.1.3 Crisp model

4.1.3.1 Case -1(Here demand falls to zero, before the deterioration set in)

Let I(t) be the inventory at any time t, $(0 \leq t \leq T)$. During $0 \leq t \leq t_0$ the inventory level I(t) decrease due to customer demand only, in $t_0 \leq t \leq t_e$ the inventory level I(t) remain same, in $t_e \leq t \leq t_d$ the inventory level decrease due to customer demand only, in $t_d \leq t \leq t_e$, the inventory level decrease due to customer demand and deterioration, and reaches to zero at $t = t_1$. During the time interval $t_1 \leq t \leq T$, the shortages with fully backlogged continued.

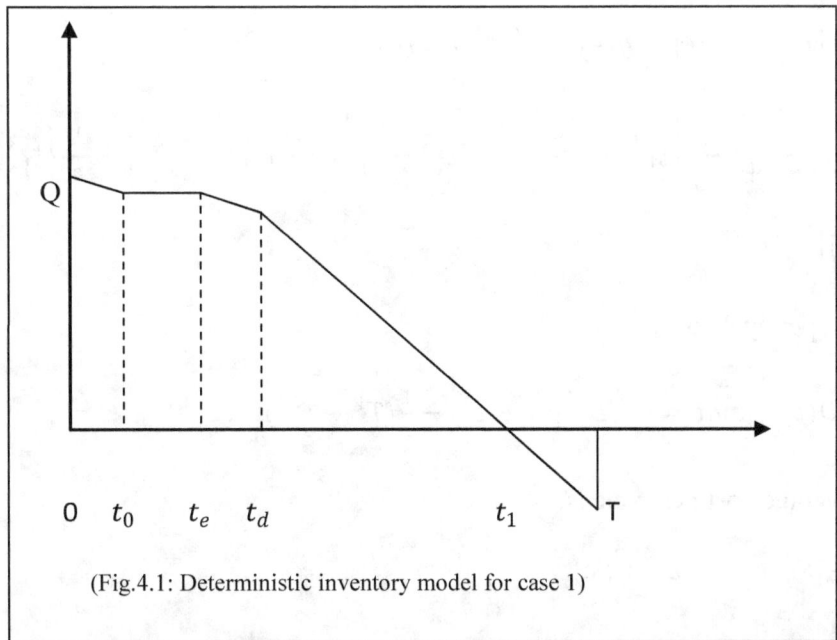

(Fig.4.1: Deterministic inventory model for case 1)

The differential equations describing the above model as follows,

$$\frac{dI(t)}{dt} = \begin{cases} -D, & 0 \leq t \leq t_0 \\ 0, & t_0 \leq t \leq t_e \\ -D, & t_e \leq t \leq t_d \\ -D - 2\theta t I(t), & t_d \leq t \leq t_1 \\ -D, & t_1 \leq t \leq T. \end{cases} \quad (4.1)$$

With boundary conditions I(t) = Q and $I(t_1) = 0$.

Corresponding solutions of the above differential equations after applying the boundary condition are,

$$I(t) = \begin{cases} Q - Dt, & 0 \leq t \leq t_0, \\ Q - Dt_0, & t_0 \leq t \leq t_e, \\ Q + D(t_e - t_0) - Dt, & t_e \leq t \leq t_d, \\ D(t_1 - t)(1 - \theta t^2) + \frac{D\theta}{3}(t_1{}^3 - t^3) & t_d \leq t \leq t_1, \\ D(t_1 - t), & t_1 \leq t \leq T. \end{cases} \quad (4.2)$$

Now the holding cost per cycle is,

$$\text{Hc} = c_h\{\int_0^{t_0} I(t)dt + \int_{t_0}^{t_e} I(t)dt + \int_{t_e}^{t_d} I(t)dt + \int_{t_d}^{t_1} I(t)dt\}$$

$$= c_h\left[\left(Qt_0 - \frac{Dt_0{}^2}{2}\right) + (Q - Dt_0)(t_e - t_0) + Q(t_d - t_e) + D(t_e - t_0)(t_d - t_e) - \frac{D}{2}(t_d{}^2 - t_e{}^2) + D\left\{t_1(t_1 - t_d) - \frac{1}{2}(t_1{}^2 - t_d{}^2) - \frac{Qt_1}{3}(t_1{}^3 - t_d{}^3) + \frac{\theta}{4}(t_1{}^4 - t_d{}^4)\right\} + \frac{D\theta}{3}\{t_1{}^3(t_1 - t_d) - \frac{1}{4}(t_1{}^4 - t_d{}^4)\}\right].$$

Deteriorating cost per cycle is,

$$\text{Dc} = c_d\int_{t_d}^{t_1} D\left\{D(t_1 - t)(1 - \theta t^2) + \frac{D\theta}{3}(t_1{}^3 - t^3)\right\}dt$$

$$= c_dD\left[D\left\{t_1(t_1 - t_d) - \frac{1}{2}(t_1{}^2 - t_d{}^2) - \frac{\theta t_1}{3}(t_1{}^3 - t_d{}^3) + \frac{\theta}{4}(t_1{}^4 - t_d{}^4)\right\} + \frac{D\theta}{3}\left\{t_1{}^3(t_1 - t_d) - \frac{1}{4}(t_1{}^4 - t_d{}^4)\right\}\right].$$

Shortages cost per cycle is,

$$\text{Sc} = -c_s\int_{t_1}^{T} D(t_1 - t)dt = -c_sD\{t_1(T - t_1) - \frac{1}{2}(T^2 - t_1{}^2)\}.$$

So the total average cost per cycle is,

$$Tac_1(t_1) = \frac{1}{T}[Fc + Sc + Dc + Sc]$$

$$= \frac{1}{T}\left[c_f + c_h\left[\left(Qt_0 - \frac{Dt_0{}^2}{2}\right) + (Q - Dt_0)(t_e - t_0) + Q(t_d - t_e) + D(t_e - t_0)(t_d - t_e) - \frac{D}{2}(t_d{}^2 - t_e{}^2) + D\left\{t_1(t_1 - t_d) - \frac{1}{2}(t_1{}^2 - t_d{}^2) - \frac{Qt_1}{3}(t_1{}^3 - t_d{}^3) + \frac{\theta}{4}(t_1{}^4 - t_d{}^4)\right\} + \frac{D\theta}{3}\left\{t_1{}^3(t_1 - t_d) - \frac{1}{4}(t_1{}^4 - t_d{}^4)\right\}\right] + c_dD\left[D\left\{t_1(t_1 - t_d) - \frac{1}{2}(t_1{}^2 - t_d{}^2) - \frac{\theta t_1}{3}(t_1{}^3 - t_d{}^3) + \frac{\theta}{4}(t_1{}^4 - t_d{}^4)\right\} + \frac{D\theta}{3}\left\{t_1{}^3(t_1 - t_d) - \frac{1}{4}(t_1{}^4 - t_d{}^4)\right\}\right] - c_sD\left\{t_1(T - t_1) - \frac{1}{2}(T^2 - t_1{}^2)\right\}\right]$$

4.1.3.2 Case -2 (Here demand falls to zero, after the deterioration set in)

Let I(t) be the inventory level at any time t, $(0 \leq t \leq T)$. During $0 \leq t \leq t_d$ the inventory level I(t) decrease due to customer demand only, in $t_d \leq t \leq t_0$ the inventory level decrease due to customer demand and deterioration, in $t_0 \le t \le t_e$ the inventory level decrease due to customer demand only, finally in $t_e \le t \le t_1$, the inventory level decrease due to customer demand and deterioration, and reaches to zero at $t = t_1$. During the time interval $t_1 \le t \le T$, the shortages with fully backlogged continued.

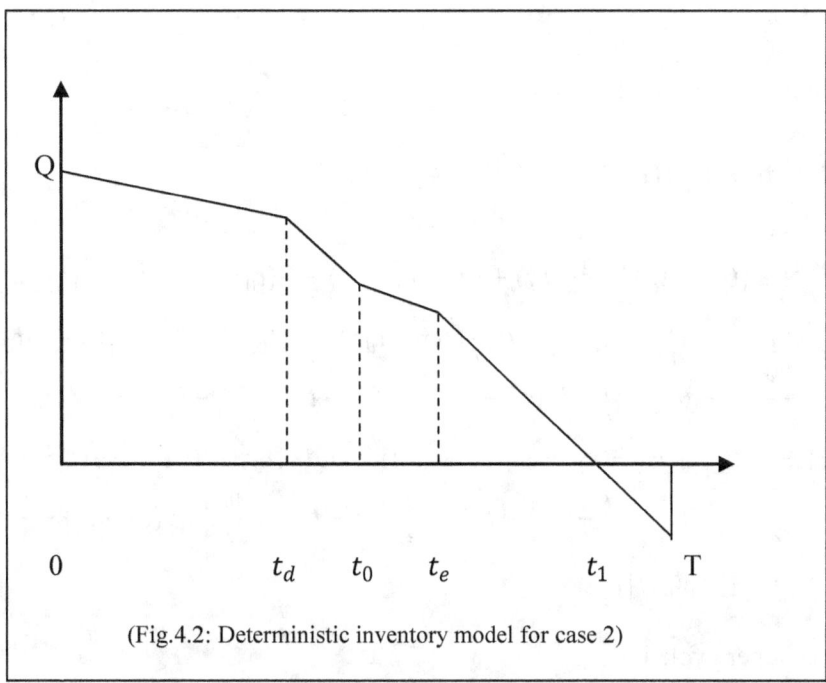

(Fig.4.2: Deterministic inventory model for case 2)

The differential equations describing the above model as follows,

$$\frac{dI(t)}{dt} = \begin{cases} -D, & 0 \le t \le t_d, \\ -D - 2\theta t I(t), & t_d \le t \le t_0, \\ -2\theta t I(t), & t_0 \le t \le t_e, \\ -D - 2\theta t I(t) & t_e \le t \le t_1, \\ -D, & t_1 \le t \le T. \end{cases} \qquad (4.3)$$

With boundary conditions $I(t) = Q$ and $I(t_1) = 0$.

Corresponding solutions of the above differential equations after applying the boundary condition are,

$$I(t) = \begin{cases} Q - Dt, & 0 \le t \le t_d, \\ (Q - Dt_d)\left(1 + \theta t_d{}^2 - \theta t^2\right) + D\left(t_d + \frac{\theta}{3}t_d{}^3 - \theta t_d t^2\right) - D\left(t - \frac{2}{3}\theta t^3\right), & t_d \le t \le t_0, \\ (Q - Dt_d)\left(1 + \theta t_d{}^2 - \theta t_0{}^2\right) + D\left(t_d + \frac{\theta}{3}t_d{}^3 - \theta t_d t_0{}^2\right) - D\left(t_0 - \frac{2}{3}\theta t_0{}^3\right) \\ \quad + (Q - Dt_d)\theta(t_0{}^2 - t^2) + D\theta t_d(t_0{}^2 - t^2) - Dt_0\theta(t_0{}^2 - t^2), & t_0 \le t \le t_e, \\ D(t_1 - t)(1 - \theta t^2) + \frac{D\theta}{3}(t_1{}^3 - t^3), & t_e \le t \le t_1, \\ D(t_1 - t), & t_1 \le t \le T. \end{cases}$$

(4.4)

Now the holding cost per cycle is,

$$\text{Hc} = c_h\{\int_0^{t_d} I(t)dt + \int_{t_d}^{t_0} I(t)d + \int_{t_0}^{t_e} I(t)d + \int_{t_e}^{t_1} I(t)d\}$$

$$= c_h\Big[Qt_d - \frac{Dt_d{}^2}{2} + (Q - Dt_d)\Big\{(t_0 - t_d) + \theta t_d{}^2(t_0 - t_d) - \frac{\theta}{3}(t_0{}^3 - t_d{}^3)\Big\} + D\Big\{t_d(t_0 - t_d)\frac{\theta}{3}t_d{}^3(t_0 - t_d) - \frac{\theta t_d}{3}(t_0{}^3 - t_d{}^3)\Big\} - D\Big\{\frac{1}{2}(t_0{}^2 - t_d{}^2) - \frac{1}{6}(t_0{}^4 - t_d{}^4)\Big\} + \Big\{(Q - Dt_d)(1 + \theta t_d{}^2 - \theta t_0{}^2) + D\Big(t_d + \frac{\theta}{3}t_d{}^3 - \theta t_d t_0{}^2\Big) - D\Big(t_0 - \frac{2\theta}{3}t_0{}^3\Big)\Big\}(t_e - t_0) + (Q - Dt_d)\theta\Big\{t_0{}^2(t_e - t_0) - \frac{1}{3}(t_e{}^3 - t_0{}^3)\Big\} + D\theta t_d\Big\{t_0{}^2(t_e - t_0) - \frac{1}{3}(t_e{}^3 - t_0{}^3)\Big\} - D\theta t_0\Big\{t_0{}^2(t_e - t_0) - \frac{1}{3}(t_e{}^3 - t_0{}^3)\Big\} + D\{t_1(t_1 - t_e) - \frac{1}{2}(t_1{}^2 - t_e{}^2) - \frac{\theta t_1}{3}(t_1{}^3 - t_e{}^3) + \frac{\theta}{4}(t_1{}^4 - t_e{}^4) + \frac{D\theta}{3}\Big\{t_1{}^3(t_1 - t_e) - \frac{1}{4}(t_1{}^4 - t_e{}^4)\Big\} + D\Big\{t_1(t_1 - t_e) - \frac{1}{2}(t_1{}^2 - t_e{}^2)\Big\}\Big].$$

Deteriorating cost per cycle is,

$$\text{Dc} = c_d\{\int_{t_d}^{t_0} DI(t)dt + \int_{t_0}^{t_e} DI(t)dt + \int_{t_e}^{t_1} DI(t)dt\}$$

$$= c_d D\Big[(Q - Dt_d)\Big\{(t_0 - t_d) + \theta t_d{}^2(t_0 - t_d) - \frac{\theta}{3}(t_0{}^3 - t_d{}^3)\Big\} + D\Big\{t_d(t_0 - t_d) + \frac{\theta}{3}t_d{}^3(t_0 - t_d) - \frac{\theta t_d}{3}(t_0{}^3 - t_d{}^3)\Big\} - D\Big\{\frac{1}{2}(t_0{}^2 - t_d{}^2) - \frac{1}{6}(t_0{}^4 - t_d{}^4)\Big\} + \Big\{(Q - Dt_d)(1 + \theta t_d{}^2 - \theta t_0{}^2) + D\Big(t_d + \frac{\theta}{3}t_d{}^3 - \theta t_d t_0{}^2\Big) - D\Big(t_0 - \frac{2\theta}{3}t_0{}^3\Big)\Big\}(t_e - t_0) + (Q - Dt_d)\theta\Big\{t_0{}^2(t_e - t_0) - \frac{1}{3}(t_e{}^3 - t_0{}^3)\Big\} + D\theta t_d\Big\{t_0{}^2(t_e - t_0) - \frac{1}{3}(t_e{}^3 - t_0{}^3)\Big\} - D\theta t_0\Big\{t_0{}^2(t_e - t_0) - \frac{1}{3}(t_e{}^3 - t_0{}^3)\Big\} + D\{t_1(t_1 - t_e) - \frac{1}{2}(t_1{}^2 - t_e{}^2) - \frac{\theta t_1}{3}(t_1{}^3 - t_e{}^3) + \frac{\theta}{4}(t_1{}^4 - t_e{}^4) + \frac{D\theta}{3}\Big\{t_1{}^3(t_1 - t_e) - \frac{1}{4}(t_1{}^4 - t_e{}^4)\Big\} + D\Big\{t_1(t_1 - t_e) - \frac{1}{2}(t_1{}^2 - t_e{}^2)\Big\}\Big].$$

Shortages cost per cycle is,

$$\text{Sc} = -c_s\int_{t_1}^{T} D(t_1 - t)dt$$

$$= -c_s D\{t_1(T - t_1) - \frac{1}{2}(T^2 - t_1{}^2)\}.$$

So the total average cost per cycle is,

$$Tac_2(t_1) = \frac{1}{T}[Fc + Sc + Dc + Sc]$$

$$= \frac{1}{T}\left[c_f + c_h\left[Qt_d - \frac{Dt_d^2}{2} + (Q - Dt_d)\left\{(t_0 - t_d) + \theta t_d^2(t_0 - t_d) - \frac{\theta}{3}(t_0^3 - t_d^3)\right\} + D\left\{t_d(t_0 - t_d) + \right.\right.\right.$$

$$\frac{\theta}{3}t_d^3(t_0 - t_d) - \frac{\theta t_d}{3}(t_0^3 - t_d^3)\right\} - D\left\{\frac{1}{2}(t_0^2 - t_d^2) - \frac{1}{6}(t_0^4 - t_d^4)\right\} + \left\{(Q - Dt_d)(1 + \theta t_d^2 - \right.$$

$$\theta t_0^2) + D\left(t_d + \frac{\theta}{3}t_d^3 - \theta t_d t_0^2\right) - D\left(t_0 - \frac{2\theta}{3}t_0^3\right)\right\}(t_e - t_0) + (Q - Dt_d)\theta\left\{t_0^2(t_e - t_0) - \right.$$

$$\frac{1}{3}(t_e^3 - t_0^3)\right\} + D\theta t_d\left\{t_0^2(t_e - t_0) - \frac{1}{3}(t_e^3 - t_0^3)\right\} - D\theta t_0\left\{t_0^2(t_e - t_0) - \frac{1}{3}(t_e^3 - t_0^3)\right\} + $$

$$D\{t_1(t_1 - t_e) - \frac{1}{2}(t_1^2 - t_e^2) - \frac{\theta t_1}{3}(t_1^3 - t_e^3) + \frac{\theta}{4}(t_1^4 - t_e^4) + \frac{D\theta}{3}\left\{t_1^3(t_1 - t_e) - \frac{1}{4}(t_1^4 - t_e^4)\right\} + $$

$$D\{t_1(t_1 - t_e) - \frac{1}{2}(t_1^2 - t_e^2)\}\right] + c_d D\left[(Q - Dt_d)\left\{(t_0 - t_d) + \theta t_d^2(t_0 - t_d) - \frac{\theta}{3}(t_0^3 - t_d^3)\right\} + $$

$$D\left\{t_d(t_0 - t_d) + \frac{\theta}{3}t_d^3(t_0 - t_d) - \frac{\theta t_d}{3}(t_0^3 - t_d^3)\right\} - D\left\{\frac{1}{2}(t_0^2 - t_d^2) - \frac{1}{6}(t_0^4 - t_d^4)\right\} + \left\{(Q - \right.$$

$$Dt_d)(1 + \theta t_d^2 - \theta t_0^2) + D\left(t_d + \frac{\theta}{3}t_d^3 - \theta t_d t_0^2\right) - D\left(t_0 - \frac{2\theta}{3}t_0^3\right)\right\}(t_e - t_0) + $$

$$(Q - Dt_d)\theta\left\{t_0^2(t_e - t_0) - \frac{1}{3}(t_e^3 - t_0^3)\right\} + D\theta t_d\left\{t_0^2(t_e - t_0) - \frac{1}{3}(t_e^3 - t_0^3)\right\} - D\theta t_0\left\{t_0^2(t_e - \right.$$

$$t_0) - \frac{1}{3}(t_e^3 - t_0^3)\right\} + D\{t_1(t_1 - t_e) - \frac{1}{2}(t_1^2 - t_e^2) - \frac{\theta t_1}{3}(t_1^3 - t_e^3) + \frac{\theta}{4}(t_1^4 - t_e^4) + \frac{D\theta}{3}\left\{t_1^3(t_1 - \right.$$

$$t_e) - \frac{1}{4}(t_1^4 - t_e^4)\right\} + D\left\{t_1(t_1 - t_e) - \frac{1}{2}(t_1^2 - t_e^2)\right\}\right]\right].$$

4.1.3.3 Case -3 (With-out account of any special condition)

Let I(t) be the inventory level at any time t, $(0 \leq t \leq T)$. During $0 \leq t \leq t_d$ the inventory level decrease due to customer demand only, in $t_d \leq t \leq t_1$ the inventory level decreases due to customer demand and deterioration, and reaches to zero at $t = t_1$. During time interval $t_1 \leq t \leq T$, the shortages with fully backlogged continued.

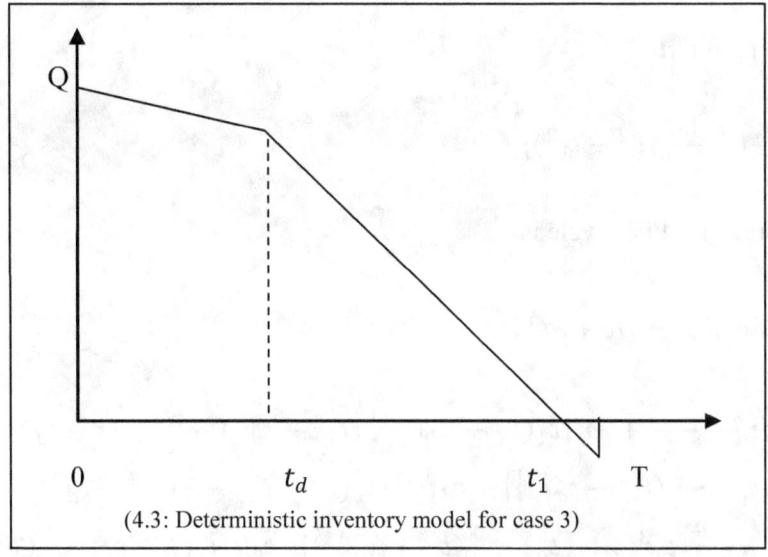

(4.3: Deterministic inventory model for case 3)

The differential equations describing the above model as follows,

$$\frac{dI(t)}{dt} = \begin{cases} -D, & 0 \le t \le t_d, \\ -D - 2\theta t I(t), & t_d \le t \le t_1, \\ -D, & t_1 \le t \le T. \end{cases} \tag{4.5}$$

With boundary conditions I(t) = Q and $I(t_1) = 0$.

Corresponding solutions of the above differential equations after applying the boundary condition are,

$$I(t) = \begin{cases} Q - Dt, & 0 \le t \le t_d, \\ D(t_1 - t)(1 - \theta t^2) + \frac{D\theta}{3}(t_1{}^3 - t^3), & t_d \le t \le t_1, \\ D(t_1 - t), & t_1 \le t \le T. \end{cases} \tag{4.6}$$

Now the holding cost per cycle is,

$$Hc = c_h\{\int_0^{t_d} I(t)dt + \int_{t_d}^{t_1} I(t)dt\}$$

$$= c_h\left[Qt_d - \frac{Dt_d{}^2}{2} + D\{t_1(t_1 - t_d) - \frac{1}{2}(t_1{}^2 - t_d{}^2) - \frac{\theta t_1}{3}(t_1{}^3 - t_d{}^3) + \frac{\theta}{4}(t_1{}^4 - t_d{}^4)\} + \frac{D\theta}{3}\{t_1{}^3(t_1 - t_d) - \frac{1}{4}(t_1{}^4 - t_d{}^4)\}\right]$$

Deteriorating cost per cycle is,

$$Dc = c_d\int_{t_d}^{t_1} DI(t)dt$$

$$= c_d\left[D^2\{t_1(t_1 - t_d) - \frac{1}{2}(t_1{}^2 - t_d{}^2) - \frac{\theta t_1}{3}(t_1{}^3 - t_d{}^3) + \frac{\theta}{4}(t_1{}^4 - t_d{}^4)\} + \frac{D^2\theta}{3}\{t_1{}^3(t_1 - t_d) - \frac{1}{4}(t_1{}^4 - t_d{}^4)\}\right].$$

Shortages cost per cycle is,

$$Sc = -c_s\int_{t_1}^{T} D(t_1 - t)dt$$

$$= -c_s D\{t_1(T - t_1) - \frac{1}{2}(T^2 - t_1{}^2)\}.$$

So the total average cost per cycle is,

$$Tac_3(t_1) = \frac{1}{T}[Fc + Sc + Dc + Sc]$$

$$= \frac{1}{T}\left[c_f + c_h\left[Qt_d - \frac{Dt_d{}^2}{2} + D\{t_1(t_1 - t_d) - \frac{1}{2}(t_1{}^2 - t_d{}^2) - \frac{\theta t_1}{3}(t_1{}^3 - t_d{}^3) + \frac{\theta}{4}(t_1{}^4 - t_d{}^4)\} + \frac{D\theta}{3}\{t_1{}^3(t_1 - t_d) - \frac{1}{4}(t_1{}^4 - t_d{}^4)\}\right] + c_d[D^2\{t_1(t_1 - t_d) - \frac{1}{2}(t_1{}^2 - t_d{}^2) - \frac{\theta t_1}{3}(t_1{}^3 - t_d{}^3) + \frac{\theta}{4}(t_1{}^4 - t_d{}^4)\} + \frac{D^2\theta}{3}\{t_1{}^3(t_1 - t_d) - \frac{1}{4}(t_1{}^4 - t_d{}^4)\}] - c_s D\{t_1(T - t_1) - \frac{1}{2}(T^2 - t_1{}^2)\}\right].$$

According to necessary and sufficient condition for minimization problem, we must have,

$$\frac{dTac_i(t_1)}{dt_1} = 0 \text{ and } \frac{d^2Tac_i(t_1)}{dt^2_1} > 0, \ i = 1,2,3.$$

4.1.4 Fuzzy model

Due to uncertainly lets us assume that, $\tilde{c}_f = (c^1{}_f, c^2{}_f, c^3{}_f), \widetilde{c_h} = (c^1{}_h, c^2{}_h, c^3{}_h), \tilde{c}_s = (c^1{}_s, c^2{}_s, c^3{}_s), \tilde{c}_d = (c^1{}_d, c^2{}_d, c^3{}_d)$, be triangular fuzzy number (TFN) then using nearest interval approximation (NIA) method, we have transformed all triangular fuzzy numbers (TFN) into interval numbers i.e., $[c_f{}^L, c_f{}^U], [c_h{}^L, c_h{}^U], [c_s{}^L, c_s{}^U], [c_d{}^L, c_d{}^U]$ and according to section 1.10 the total average cost is given by,

4.1.4.1 For case -1

$$\widetilde{Tac_1(t_1)} = \frac{1}{T}\left[(c_f{}^L)^{1-s}(c_f{}^U)^s + (c_h{}^L)^{1-s}(c_h{}^U)^s\left[\left(Qt_0 - \frac{Dt_0{}^2}{2}\right) + (Q - Dt_0)(t_e - t_0) + Q(t_d - t_e) + D(t_e - t_0)(t_d - t_e) - \frac{D}{2}(t_d{}^2 - t_e{}^2) + D\left\{t_1(t_1 - t_d) - \frac{1}{2}(t_1{}^2 - t_d{}^2) - \frac{Qt_1}{3}(t_1{}^3 - t_d{}^3) + \frac{\theta}{4}(t_1{}^4 - t_d{}^4)\right\} + \frac{D\theta}{3}\left\{t_1{}^3(t_1 - t_d) - \frac{1}{4}(t_1{}^4 - t_d{}^4)\right\}\right] + (c_d{}^L)^{1-s}(c_d{}^U)^s D\left[D\left\{t_1(t_1 - t_d) - \frac{1}{2}(t_1{}^2 - t_d{}^2) - \frac{\theta t_1}{3}(t_1{}^3 - t_d{}^3) + \frac{\theta}{4}(t_1{}^4 - t_d{}^4)\right\} + \frac{D\theta}{3}\left\{t_1{}^3(t_1 - t_d) - \frac{1}{4}(t_1{}^4 - t_d{}^4)\right\}\right] - (c_s{}^L)^{1-s}(c_s{}^U)^s D\{t_1(T - t_1) - \frac{1}{2}(T^2 - t_1{}^2)\}\right].$$

4.1.4.2 For case -2

$$\widetilde{Tac_2(t_1)} = \frac{1}{T}\left[(c_f{}^L)^{1-s}(c_f{}^U)^s + (c_h{}^L)^{1-s}(c_h{}^U)^s\left[Qt_d - \frac{Dt_d{}^2}{2} + (Q - Dt_d)\left\{(t_0 - t_d) + \theta t_d{}^2(t_0 - t_d) - \frac{\theta}{3}(t_0{}^3 - t_d{}^3)\right\} + D\left\{t_d(t_0 - t_d) + \frac{\theta}{3}t_d{}^3(t_0 - t_d) - \frac{\theta t_d}{3}(t_0{}^3 - t_d{}^3)\right\} - D\left\{\frac{1}{2}(t_0{}^2 - t_d{}^2) - \frac{1}{6}(t_0{}^4 - t_d{}^4)\right\} + \left\{(Q - Dt_d)(1 + \theta t_d{}^2 - \theta t_0{}^2) + D\left(t_d + \frac{\theta}{3}t_d{}^3 - \theta t_d t_0{}^2\right) - D\left(t_0 - \frac{2\theta}{3}t_0{}^3\right)\right\}(t_e - t_0) + (Q - Dt_d)\theta\left\{t_0{}^2(t_e - t_0) - \frac{1}{3}(t_e{}^3 - t_0{}^3)\right\} + D\theta t_d\left\{t_0{}^2(t_e - t_0) - \frac{1}{3}(t_e{}^3 - t_0{}^3)\right\} - D\theta t_0\left\{t_0{}^2(t_e - t_0) - \frac{1}{3}(t_e{}^3 - t_0{}^3)\right\} + D\{t_1(t_1 - t_e) - \frac{1}{2}(t_1{}^2 - t_e{}^2) - \frac{\theta t_1}{3}(t_1{}^3 - t_e{}^3) + \frac{\theta}{4}(t_1{}^4 - t_e{}^4) + \frac{D\theta}{3}\left\{t_1{}^3(t_1 - t_e) - \frac{1}{4}(t_1{}^4 - t_e{}^4)\right\} + D\{t_1(t_1 - t_e) - \frac{1}{2}(t_1{}^2 - t_e{}^2)\}\right] + (c_d{}^L)^{1-s}(c_d{}^U)^s D\left[(Q - Dt_d)\left\{(t_0 - t_d) + \theta t_d{}^2(t_0 - t_d) - \frac{\theta}{3}(t_0{}^3 - t_d{}^3)\right\} + D\left\{t_d(t_0 - t_d) + \frac{\theta}{3}t_d{}^3(t_0 - t_d) - \frac{\theta t_d}{3}(t_0{}^3 - t_d{}^3)\right\} - D\left\{\frac{1}{2}(t_0{}^2 - t_d{}^2) - \frac{1}{6}(t_0{}^4 - t_d{}^4)\right\} + \left\{(Q - Dt_d)(1 + \theta t_d{}^2 - \theta t_0{}^2) + D\left(t_d + \frac{\theta}{3}t_d{}^3 - \theta t_d t_0{}^2\right) - D\left(t_0 - \frac{2\theta}{3}t_0{}^3\right)\right\}(t_e - t_0) + (Q - Dt_d)\theta\left\{t_0{}^2(t_e - t_0) - \frac{1}{3}(t_e{}^3 - t_0{}^3)\right\} + D\theta t_d\left\{t_0{}^2(t_e - t_0) - \frac{1}{3}(t_e{}^3 - t_0{}^3)\right\} - D\theta t_0\left\{t_0{}^2(t_e - t_0) - \frac{1}{3}(t_e{}^3 - t_0{}^3)\right\} + D\{t_1(t_1 - t_e) - \frac{1}{2}(t_1{}^2 - t_e{}^2) - \frac{\theta t_1}{3}(t_1{}^3 - t_e{}^3) + \frac{\theta}{4}(t_1{}^4 - t_e{}^4) + \frac{D\theta}{3}\left\{t_1{}^3(t_1 - t_e) - \frac{1}{4}(t_1{}^4 - t_e{}^4)\right\} - (c_s{}^L)^{1-s}(c_s{}^U)^s D\left\{t_1(t_1 - t_e) - \frac{1}{2}(t_1{}^2 - t_e{}^2)\right\}\right]\right].$$

4.1.4.3 For case -3

$$\widehat{Tac_3(t_1)} = \frac{1}{T}\Big[(c_s{}^L)^{1-s}(c_s{}^U)^s + (c_h{}^L)^{1-s}(c_h{}^U)^s\Big[Qt_d - \frac{Dt_d{}^2}{2} + D\Big\{t_1(t_1 - t_d) - \frac{1}{2}(t_1{}^2 - t_d{}^2) -$$

$$\frac{\theta t_1}{3}(t_1{}^3 - t_d{}^3) + \frac{\theta}{4}(t_1{}^4 - t_d{}^4)\Big\} + \frac{D\theta}{3}\Big\{t_1{}^3(t_1 - t_d) - \frac{1}{4}(t_1{}^4 - t_d{}^4)\Big\}\Big] + (c_d{}^L)^{1-s}(c_d{}^U)^s[D^2\{t_1(t_1 - t_d - $$

$$\frac{1}{2}(t_1{}^2 - t_d{}^2) - \frac{\theta t_1}{3}(t_1{}^3 - t_d{}^3) + \frac{\theta}{4}(t_1{}^4 - t_d{}^4)\} + \frac{D^2\theta}{3}\{t_1{}^3(t_1 - t_d) - \frac{1}{4}(t_1{}^4 - t_d{}^4)\}] - $$

$$(c_s{}^L)^{1-s}(c_s{}^U)^s D\{t_1(T - t_1) - \frac{1}{2}(T^2 - t_1{}^2)\}\Big].$$

According to necessary and sufficient condition for minimization problem, we must have,

$$\frac{d\widehat{Tac}_i(t_1)}{dt_1} = 0 \text{ and } \frac{d^2\widehat{Tac}_i(t_1)}{dt^2{}_1} > 0, \ i = 1,2,3.$$

4.2   Numerical Solution

Case-1: For crisp solutions, let us take the in-put values are,

(Table No. 4.1: Crisp in put values)

c_f	c_h	c_d	c_s	D	Q	t_0	t_e	t_d	θ	T
100	60	20	10	6	200	0.1	0.2	0.4	0.1	10

Then the out-put values are,

(Table No. 4.2: Crisp out-put values)

$t_1{}^*$	$Tac(t_1{}^*)$
0.843	751.237

For fuzzy solution:

When the input data's of inventory model are taken as triangular fuzzy numbers i.e., $\widetilde{c_f} = (80,100,120), \widetilde{c_h} = (50, 60, 70), \widetilde{c_d} = (16, 20, 24), \widetilde{c_s} = (8, 10, 12)$, and others input values are same as table-4.1. Using nearest interval approximation method, we get the corresponding interval numbers and interval-valued functions, as follows

$$c_f = [90,110], \Rightarrow \widehat{c_f} = (90)^{1-s}(110)^s \in [90,110],$$

$$c_h = [55, 65], \Rightarrow \widehat{c_h} = (55)^{1-s}(65)^s \in [55, 65],$$

$$c_d = [18, 22], \Rightarrow \widehat{c_d} = (18)^{1-s}(22)^s \in [18, 22],$$

$c_s = [9, 11], \Rightarrow \hat{c}_s = (9)^{1-s}(11)^s \in [9, 11],$ where s $\in$ [0,1].

The optimal solution of the fuzzy model by interval-valued parametric geometric programming is presented in Table 4.3.

(Table- 4.3.Optimal Solution for Fuzzy Inventory Model)

s	$t_1{}^*$	$Tac(t_1{}^*)$
0.0	0.840	684.154
0.2	0.841	709.086
0.4	0.842	734.934
0.6	0.843	761.731
0.8	0.844	789.514
1.0	0.845	818.319

Here we have given a rough sketch, which shown how change the value of $Tac^*(S^*, D^*, q^*)$ for different values of s.

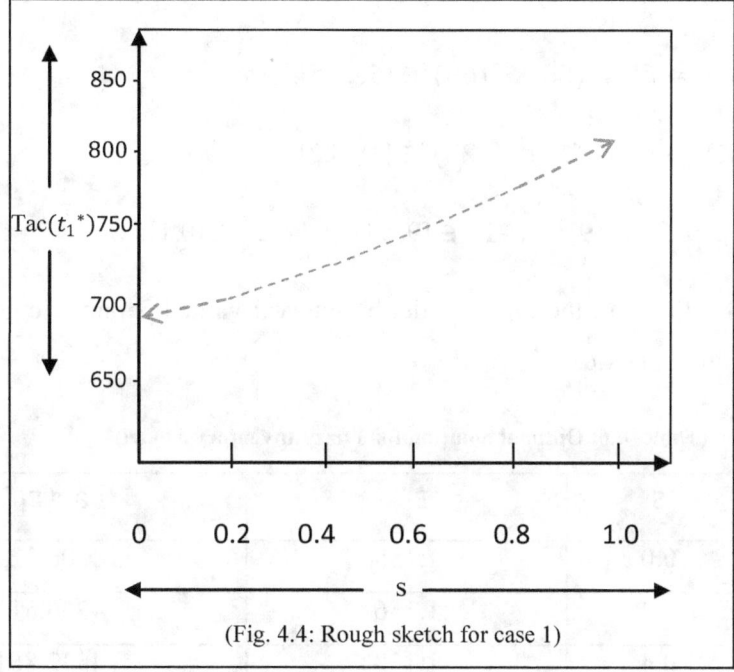

(Fig. 4.4: Rough sketch for case 1)

Case-2: For crisp solutions, let us take the in-put values are

(Table No. 4.4: Crisp in put values)

c_f	c_h	c_d	c_s	D	Q	t_0	t_e	t_d	θ	T

100	60	20	10	6	200	0.6	0.8	0.4	0.1	10

Then the out-put values are

(Table No. 4.5: Crisp out- put values)

$t_1{}^*$	$\mathrm{Tac}(t_1{}^*)$
1.158	4638.772

For fuzzy solution:

When the input data's of inventory model are taken as triangular fuzzy numbers i.e., $\tilde{c}_f =$ (80,100,120),$\widetilde{c_h}$ = (50, 60, 70), $\widetilde{c_d}$= (16, 20, 24),$\tilde{c}_s$ = (8, 10, 12) and others input values are same as table-4.1.Using nearest interval approximation method, we get the corresponding interval numbers and interval-valued functions, as follows

$$c_f = [90,110], \Rightarrow \widehat{c_f} = (90)^{1-s}(110)^s \in [90,110],$$

$$c_h = [55, 65], \Rightarrow \widehat{c_h} = (55)^{1-s}(65)^s \in [55, 65],$$

$$c_d = [18, 22], \Rightarrow \widehat{c_d} = (18)^{1-s}(22)^s \in [18, 22],$$

$$c_s = [9, 11], \Rightarrow \widehat{c_s} = (9)^{1-s}(11)^s \in [9, 11], \text{ where s} \in [0,1].$$

The optimal solution of the fuzzy model by interval-valued parametric geometric programming is presented in Table 4.6.

(Table- 4.6: Optimal Solution for Fuzzy Inventory Model)

s	$t_1{}^*$	$\mathrm{Tac}(t_1{}^*)$
0.0	1.155	4206.323
0.2	1.156	4366.517
0.4	1.158	4532.861
0.6	1.159	4705.593
0.8	1.160	4884.960
1.0	1.161	5071.220

Here we have given a rough sketch, which shown how change the value of $\text{Tac}^*(S^*, D^*, q^*)$ for different values of s.

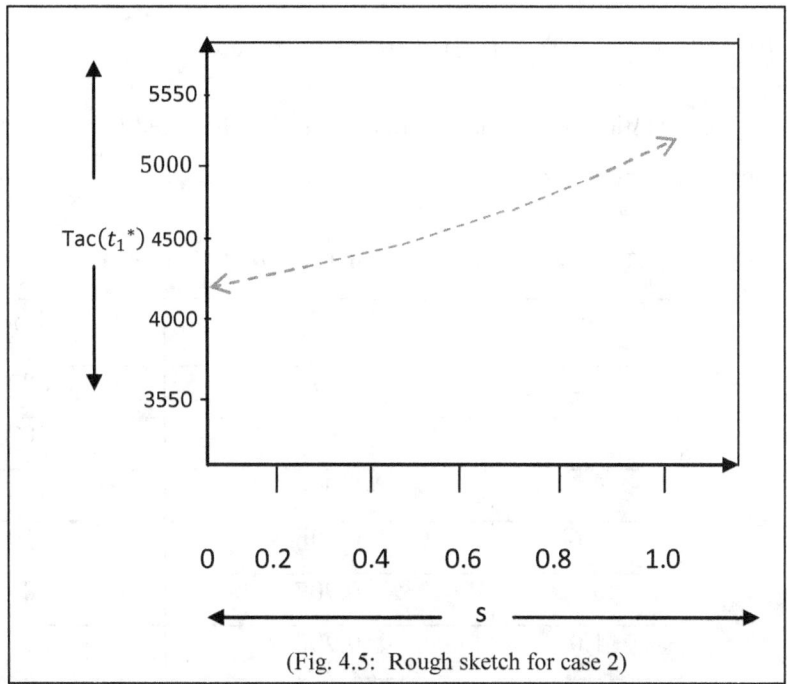

(Fig. 4.5: Rough sketch for case 2)

Case-3: For crisp solutions, let us take the in-put value;

(Table No. 4.7: Crisp in put values)

c_f	c_h	c_d	c_s	D	Q	t_d	θ	T
100	60	20	10	6	200	0.4	0.1	10

Then the out-put values are

(Table No. 4.8: Crisp out-put values)

t_1^*	$\text{Tac}(t_1^*)$
0.905	749.048

For fuzzy solutions:

When the input data's of inventory model are taken as triangular fuzzy numbers i.e., $\tilde{c}_f = (80,100,120), \widetilde{c_h} = (50, 60, 70), \widetilde{c_d} = (16, 20, 24), \tilde{c}_s = (8, 10, 12)$ and others input values are same as table-4.1. Using nearest interval approximation method, we get the corresponding interval numbers and interval-valued functions, as follows

$c_f = [90,110], \Rightarrow \hat{c}_f = (90)^{1-s}(110)^s \in [90,110],$

$c_h = [55, 65], \Rightarrow \widehat{c_h} = (55)^{1-s}(65)^s \in [55, 65],$

$c_d = [18, 22], \Rightarrow \widehat{c_d} = (18)^{1-s}(22)^s \in [18, 22],$

$c_s = [9, 11], \Rightarrow \widehat{c_s} = (9)^{1-s}(11)^s \in [9, 11],$ where s $\in [0,1]$.

The optimal solution of the fuzzy model by interval-valued parametric geometric programming is presented in Table 4.9.

(Table- 4.9: Optimal Solution for Fuzzy Inventory Model)

s	t_1^*	$Tac(t_1^*)$
0.0	0902	682.172
0.2	0.903	707.027
0.4	0.904	732.796
0.6	0.906	759.511
0.8	0.907	787.208
1.0	0.908	815.924

Here we have given a rough sketch, which shown how change the value of $Tac^*(S^*, D^*, q^*)$ for different values of s.

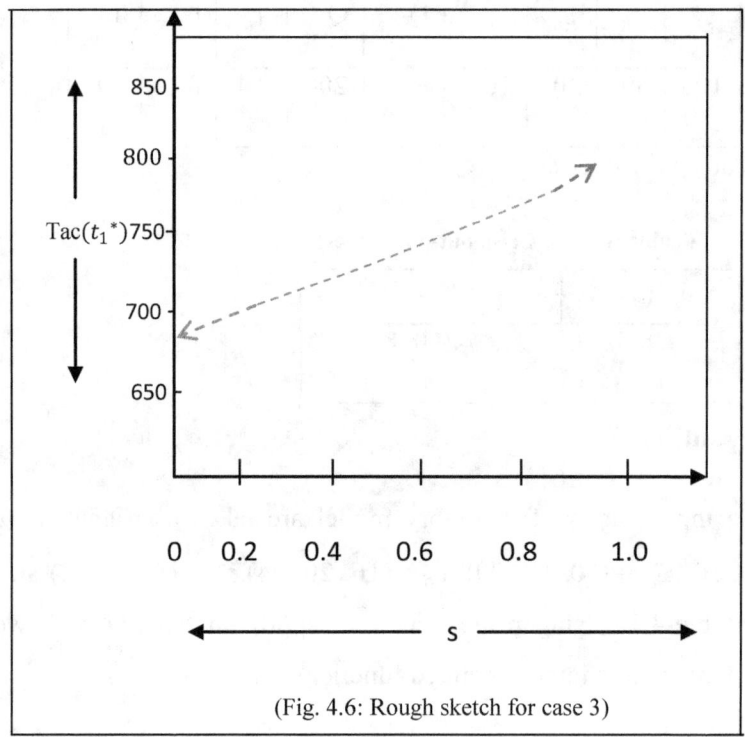

(Fig. 4.6: Rough sketch for case 3)

4.3   Conclusion

In this chapter, we have proposed a real life inventory problem in a crisp and fuzzy environment. The inventory model is developed with constant demand and shortages under fully backlogged. Here we have considered an unexpected condition (flood, strike, earthquake, etc.) when demand falls to zero and considered three cases. From numerical solution we have seen that $Tac_3(t_1) \leq Tac_1(t_1) \leq Tac_2(t_1)$ and it conclude that the total average cost is maximum when demand falls to zero after deterioration start, and minimum when no such unexpected condition arise.

Here we have considered triangular fuzzy number (TFN) for the cost parameters and use nearest interval approximation (NIA) method to convert a triangular fuzzy number (TFN) to an interval number, thereafter transformed this interval number to a parametric interval-valued functional form and solved. In future, the other type of membership functions such as piecewise linear hyperbolic, L-R fuzzy number, trapezoidal fuzzy number (TrFN) etc. can be considered for the cost parameters of the model and the model can be easily solved.

Chapter5.

A Fuzzy Inventory Model with Unit Production Cost, Time Depended Holding Cost, Without Shortages under a Space Constraint: A Geometric Programming Approach

Economic order quantity (EOQ) is a simple mathematical model to deal with inventory management (IM). It's make up a significant part of the production management. Usually an inventory model is characterized by several parameters such as cost coefficients, demand, deteriorations etc. Considering these crisp parameters as fuzzy parameters is more realistic.

*In this chapter, **an inventory model with unit production cost, time dependent holding cost, without shortages is formulated and solved.** We have considered here a single objective inventory model. In most real world situation, the objective and constraint function are imprecise in nature, hence the coefficients are imposed here in fuzzy parameters. Geometric programming (GP) provides a powerful tool for solving a variety of non-linear optimization problem. Here we have used the nearest interval approximation (NIA) technique to convert a triangular fuzzy number (TFN) to an interval number then transform this interval number to a parametric interval-valued functional form and solve the parametric problem by geometric programming (GP) technique. Here two necessary theorems have been derived. Numerical example is given to illustrate the model through this geometric programming (GP) method.*

5.1 Mathematical Model

An EOQ model is developed under the following notations and assumptions.

5.1.1 Notations

I(t):Inventory level at any time, $t \geq 0$.

D: Constant demand per unit time..

T: Cycle of length.

S: Set-up cost per batch.

H(t): Time dependent holding cost per unit per unit time.

P: Unit demand and set-up cost dependent production cost.

q: Production quantity per batch.

f(D,S): Unit production cost per cycle.

TAC(D,S,q):Total average cost per unit time.

w_0: Space area per unit quantity.

W: Total storage space area.

5.1.2 Assumptions

a) The inventory system involves only one item.

b) The replenishment occurs instantaneously at infinite rate.

c) The lead time is negligible.

d) Demand rate is constant.

e) The unit production cost is continuous function of demand and set-up cost and take the form: $P = \theta D^{-x} S^{-1}$, $\theta, x \in \mathbb{R}$ (>0).

f) Holding cost is time depended, $H(t) = H.at$,   a, H$\in \mathbb{R}$ (>0).

5.1.3 Crisp model

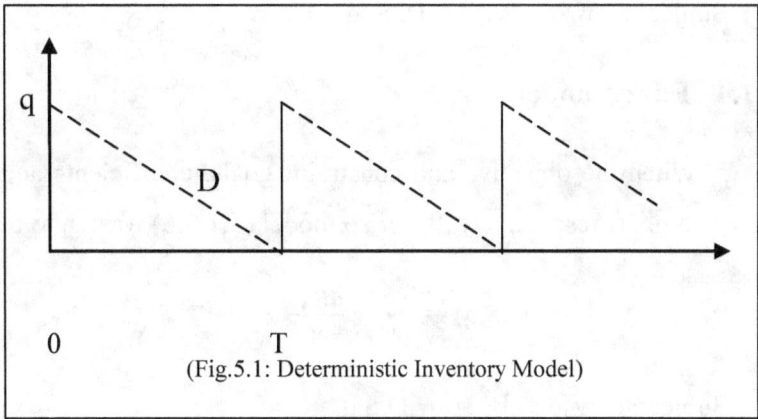

(Fig.5.1: Deterministic Inventory Model)

The differential equations describing the above model as follows,

$$\frac{dI(t)}{dt} = -D \ , \qquad 0 \leq t \leq T \qquad\qquad (5.1.3.1)$$

With the boundary condition $I(0) = q$, $I(T) = 0$.

The solution of (5.1.3.1) is obtained as

$$I(t) = q - Dt \qquad (5.1.3.2)$$

Also there are $T = q/D$.

Here inventory holding cost $= H\int_0^T at.\, I(t)dt = \frac{aHq^3}{6D^2}.$ (5.1.3.3)

Total inventory related cost per cycle = set-up cost + holding cost + production cost

$$= S + \frac{aHq^3}{6D^2} + Pq. \qquad (5.1.3.4)$$

So total average cost per cycle is given by

$$TAC(D,S,q) = \frac{1}{T}\left(S + \frac{aHq^3}{6D^2} + Pq\right)$$

$$= \frac{SD}{q} + \frac{aHq^2}{6D} + \theta D^{1-x}S^{-1}. \qquad (5.1.3.5)$$

And storage area $= w_0q$.

Hence the inventory model can be written as

$$\text{Min} \qquad TAC(D,S,q) = \frac{SD}{q} + \frac{aHq^2}{6D} + \theta D^{1-x}S^{-1}. \qquad (5.1.3.6)$$

Subject to $w_0q \leq W$, $\qquad D, S, q > 0$.

5.1.4 Fuzzy model

When the objective and constraint goals, coefficients and exponents become fuzzy sets and fuzzy numbers respectively, the crisp model (5.1.3.6) written to be a fuzzy model, as

$$\widetilde{Min} \qquad TAC(D,S,q) = \frac{SD}{q} + \frac{\tilde{a}\tilde{H}q^2}{6D} + \tilde{\theta}D^{1-x}S^{-1}$$

Subject to $w_0q \lesssim \widetilde{W}$, $\qquad D,S,q > 0$. (5.1.4.1)

5.2   Solution Procedure of Crisp Model by Geometric Programming (GP) Technique

Here the primal problem is

Min $\quad$ TAC(D,S,q) $= \frac{SD}{q} + \frac{\tilde{a}\tilde{H}q^2}{6D} + \theta D^{1-x}S^{-1}$ $\qquad$ (5.2.1)

Subject to $\quad w_0 q \leq W, \qquad$ D, S, q > 0.

Corresponding dual form of (5.2.1.1) is given by

Max d(δ) $= (\frac{1}{\delta_1})^{\delta_1} (\frac{aH}{6\delta_2})^{\delta_2} (\frac{\theta}{\delta_3})^{\delta_3} (\frac{w_0}{W\delta_{01}})^{\delta_{01}} \delta_{01}{}^{\delta_{01}}$

Subject to

$$\delta_1 + \delta_2 + \delta_3 = 1 \qquad (5.2.2)$$

$$\delta_1 - \delta_3 = 0$$

$$\delta_1 - \delta + (1-x)\delta_3 = 0$$

$$-\delta_1 + 2\delta_2 + \delta_{01} = 0$$

$$\delta_1, \delta_2, \delta_3, \delta_{01} > 0.$$

From (5.2.2) we get $\quad \delta_1 = \frac{1}{4-x}, \delta_2 = \frac{2-x}{4-x}, \delta_3 = \frac{1}{4-x}$, and $\delta_{01} = \frac{2x-3}{4-x}$.

Putting the values in the objective function (5.2.2) we get the optimal objective value of the dual problem. The values of D, S, q is obtained from the primal dual relation.

From primal dual relation we get

$$\frac{SD}{q} = \delta_1{}^* d^*(\delta),$$

$$\frac{aHq^2}{6D} = \delta_2{}^* d^*(\delta),$$

$$\theta D^{1-x}S^{-1} = \delta_3{}^* d^*(\delta),$$

$$\frac{w_0 q}{W} = 1.$$

The optimal solution of the model through the parametric approach is given by

$$S^* = \frac{6\delta_1{}^* \delta_2{}^* d^*(\delta)^2}{aH},$$

$$D^* = \frac{aHq^2}{6\delta_2{}^* d^*(\delta)},$$

$$q^* = \frac{W}{w_0}.$$

Where $d^*(\delta) = (4-x)^{\frac{1}{4-x}} \left(\frac{aH(4-x)}{(2-x)6}\right)^{\frac{2-x}{4-x}} (\theta(4-x))^{\frac{1}{4-x}} \times \left(\frac{w_0(4-x)}{W(2x-3)}\right)^{\frac{2x-3}{4-x}} (\frac{2x-3}{4-x})^{\frac{2x-3}{4-x}}.$

5.3   Geometric Programming with Fuzzy Coefficient

When all coefficients of the problem are taken as triangular fuzzy number, then the geometric programming problem is of the form

Min $\tilde{g}_0(x)$ (5.3.1)

Subject to $\tilde{g}_i(x) \lesssim 1$ $(1 \le i \le n)$

    $x > 0,$

Its objective function is $\tilde{g}_0(x) = \sum_{k=1}^{T_0} \tilde{c}_{0k} \prod_{j=1}^{m} x_j^{\alpha_{0kj}}$, and constraints of the form $\tilde{g}_i(x) = \sum_{k=1}^{T_i} \tilde{c}_{ik} \prod_{j=1}^{m} x_j^{\alpha_{ikj}}$ $(0 \le i \le n)$, are all posynomials of x in which coefficients $\tilde{c}_{0k}$ and indexes $\tilde{c}_{ik}$ are fuzzy numbers.

Here $\tilde{c}_{0k} = (c^1_{0k}, c^2_{0k}, c^3_{0k})$ and $\tilde{c}_{ik} = (c^1_{ik}, c^2_{ik}, c^3_{ik})$. Using nearest interval approximation (NIA) method, we transform all triangular fuzzy number into interval number i.e., $[c_{0k}^L, c_{0k}^U]$ and $[c_{ik}^L, c_{ik}^U]$. The geometric programming problem with imprecise parameters is of the following form

Min $\hat{g}_0(x)$ (5.3.2)

Subject to $\hat{g}_i(x) \lesssim 1$ $(1 \le i \le n)$

    $x > 0,$

Its objective function is $\hat{g}_0(x) = \sum_{k=1}^{T_0} \hat{c}_{0k} \prod_{j=1}^{m} x_j^{\alpha_{0kj}}$, and constraints of the form $\hat{g}_i(x) = \sum_{k=1}^{T_i} \hat{c}_{ik} \prod_{j=1}^{m} x_j^{\alpha_{ikj}}$ $(0 \le i \le n)$. Where $\hat{c}_{0k}$ and $\hat{c}_{ik}$ denote the interval counterparts i.e. $\hat{c}_{0k} \in [c_{0k}^L, c_{0k}^U]$ and $\hat{c}_{ik} \in [c_{ik}^L, c_{ik}^U]$. $c_{0k}^L > 0, c_{ik}^L > 0$ for all i and k. Using parametric interval-valued functional form, the problem (5.3.2) reduces to

Min $g_0(x,s) = \sum_{k=1}^{T_0} (c_{0k}^L)^{1-s} (c_{0k}^U)^s \prod_{j=1}^{m} x_j^{\alpha_{0kj}}$ (5.3.3)

Subject to $g_i(x,s) = \sum_{k=1}^{T_i} (c_{ik}^L)^{1-s} (c_{ik}^U)^s \prod_{j=1}^{m} x_j^{\alpha_{ikj}} \le 1$

$x_j > 0$ for $i = 1, 2, \ldots \ldots n$, $j = 1, 2, \ldots \ldots m$.

This is a parametric geometric programming problem. We get different solutions of this problem for different values of the parameter "s".

The dual programming of (5.3.3) is as follows

$$\text{Max} \quad d(\delta, s) = \prod_{i=0}^{n} \prod_{k=0}^{T_i} \left(\frac{(c_{ik}{}^L)^{1-s} (c_{ik}{}^U)^s}{\delta_{ik}} \right)^{\delta_{ik}} \left(\sum_{s=1}^{T_i} \delta_{is} \right)^{\left(\sum_{s=1}^{T_i} \delta_{is} \right)} \tag{5.3.4}$$

Subject to

$$\sum_{k=1}^{T_0} \delta_{ok} = 1,$$

$$\sum_{i=0}^{n} \sum_{k=1}^{T_0} \alpha_{ikj} \delta_{ik} = 0,$$

$$\delta_{ik} > 0.$$

Theorem 5.3(a)

If x is a feasible vector for the constraints PGP and if δ is a feasible vector for the corresponding DP, then $g_0(x, s) \geq d(\delta, s)$ (Primal- Dual Inequality).

Proof

The expression for $g_0(x, s)$ can be written as

$$g_0(x, s) = \sum_{k=1}^{T_0} \delta_{0k} \left(\frac{(c_{0k}{}^L)^{1-s} (c_{0k}{}^U)^s \prod_{j=1}^{m} x_j{}^{\alpha_{0kj}}}{\delta_{0k}} \right).$$

Here the weights are $\delta_{01}, \delta_{02}, \ldots \ldots \ldots, \delta_{0T_0}$ and positive terms are $\dfrac{(c_{01}{}^L)^{1-s} (c_{01}{}^U)^s \prod_{j=1}^{m} x_j{}^{\alpha_{01j}}}{\delta_{01}}$,

$$\frac{(c_{02}{}^L)^{1-s} (c_{02}{}^U)^s \prod_{j=1}^{m} x_j{}^{\alpha_{02j}}}{\delta_{02}}, \ldots \ldots \ldots, \frac{(c_{0T_0}{}^L)^{1-s} (c_{0T_0}{}^U)^s \prod_{j=1}^{m} x_j{}^{\alpha_{0T_0j}}}{\delta_{0T_0}}.$$

Now applying A.M.-.G.M inequality, we get

$$\left(\frac{(c_{01}{}^L)^{1-s} (c_{01}{}^U)^s \prod_{j=1}^{m} x_j{}^{\alpha_{01j}} + (c_{02}{}^L)^{1-s} (c_{02}{}^U)^s \prod_{j=1}^{m} x_j{}^{\alpha_{02j}} + \ldots + (c_{0T_0}{}^L)^{1-s} (c_{0T_0}{}^U)^s \prod_{j=1}^{m} x_j{}^{\alpha_{0T_0j}}}{(\delta_{01} + \delta_{02} + \cdots + \delta_{0T_0})} \right)^{(\delta_{01} + \delta_{02} + \cdots + \delta_{0T_0})}$$

$$\geq \left(\left(\frac{(c_{01}{}^L)^{1-s} (c_{01}{}^U)^s \prod_{j=1}^{m} x_j{}^{\alpha_{01j}}}{\delta_{01}} \right)^{\delta_{01}} \left(\frac{(c_{02}{}^L)^{1-s} (c_{02}{}^U)^s \prod_{j=1}^{m} x_j{}^{\alpha_{02j}}}{\delta_{02}} \right)^{\delta_{01}} \ldots \left(\frac{(c_{0T_0}{}^L)^{1-s} (c_{0T_0}{}^U)^s \prod_{j=1}^{m} x_j{}^{\alpha_{0T_0j}}}{\delta_{0T_0}} \right)^{\delta_{0T_0}} \right)$$

Or $\left(\dfrac{g_0(x,s)}{\sum_{k=1}^{T_0}\delta_{ik}}\right)^{\sum_{k=1}^{T_0}\delta_{0k}} \geq \prod_{k=1}^{T_0}\left(\dfrac{(c_{0k}{}^L)^{1-s}(c_{0k}{}^U)^s\prod_{j=1}^m x_j{}^{\alpha_{0kj}}}{\delta_{0k}}\right)^{\delta_{0k}}$ $\quad$ [as $\sum_{k=1}^{T_0}\delta_{0k}=1$]

Or $\quad g_0(x,s) \geq \left(\dfrac{(c_{0k}{}^L)^{1-s}(c_{0k}{}^U)^s}{\delta_{0k}}\right)^{\sum_{k=1}^{T_0}\delta_{ok}}\prod_{j=1}^m x_j{}^{\sum_{k=1}^{T_0}\alpha_{0kj}\,\delta_{ok}}$

Or $\quad g_0(x,s) \geq \prod_{k=1}^{T_i}\left(\dfrac{(c_{ik}{}^L)^{1-s}(c_{ik}{}^U)^s}{\delta_{ik}}\right)^{\delta_{ik}}\prod_{j=1}^m x_j{}^{\sum_{k=1}^{T_0}\alpha_{0kj}\,\delta_{ok}}$ $\qquad$ (5.3.5)

Again $g_i(x, s)$ can be written as

$$g_i(x,s) = \sum_{k=1}^{T_l}\delta_{ik}\left(\dfrac{(c_{ik}{}^L)^{1-s}(c_{ik}{}^U)^s\prod_{j=1}^m x_j{}^{\alpha_{ikj}}}{\delta_{ik}}\right).$$

Now applying A.M.-.G.M inequality, we get

Or $\quad\left(\dfrac{g_i(x,s)}{\sum_{k=1}^{T_i}\delta_{ik}}\right)^{\sum_{k=1}^{T_i}\delta_{ik}} \geq \prod_{k=1}^{T_i}\left(\dfrac{(c_{ik}{}^L)^{1-s}(c_{ik}{}^U)^s\prod_{j=1}^m x_j{}^{\alpha_{ikj}}}{\delta_{ik}}\right)^{\delta_{ik}}$

Or $\quad g_i(x,s)^{\sum_{k=1}^{T_i}\delta_{ik}} \geq \prod_{k=1}^{T_i}\left(\dfrac{(c_{ik}{}^L)^{1-s}(c_{ik}{}^U)^s}{\delta_{ik}}\right)^{\delta_{ik}}\prod_{j=1}^m x_j{}^{\sum_{k=1}^{T_i}\alpha_{ikj}\,\delta_{ik}}\left(\sum_{k=1}^{T_i}\delta_{ik}\right)^{\left(\sum_{k=1}^{T_i}\delta_{ik}\right)}$ $\quad$ (5.3.6)

Using $g_i(x,s)^{\sum_{k=1}^{T_i}\delta_{ik}} \leq 1$, we get $\qquad$ [as $g_i(x,s) \leq 1$]

$$1 \geq \prod_{k=1}^{T_i}\left(\dfrac{(c_{ik}{}^L)^{1-s}(c_{ik}{}^U)^s}{\delta_{ik}}\right)^{\delta_{ik}}\prod_{j=1}^m x_j{}^{\sum_{k=1}^{T_i}\alpha_{ikj}\,\delta_{ik}}\left(\sum_{k=1}^{T_i}\delta_{is}\right)^{\sum_{k=1}^{T_i}\delta_{ik}} \qquad (5.3.7)$$

Multiplying (5.3.5) and (5.3.7) we get

$$g_0(x,s) \geq \prod_{i=0}^n\prod_{k=0}^{T_i}\left(\dfrac{(c_{ik}{}^L)^{1-s}(c_{ik}{}^U)^s}{\delta_{ik}}\right)^{\delta_{ik}}\left(\sum_{k=1}^{T_i}\delta_{ik}\right)^{\left(\sum_{k=1}^{T_i}\delta_{ik}\right)}\prod_{j=1}^m x_j{}^{\sum_{i=1}^n\sum_{k=1}^{T_0}\alpha_{ikj}\,\delta_{ik}} \qquad (5.3.8)$$

Using orthogonal condition the inequality (5.3.8) becomes

$$g_0(x,s) \geq \prod_{i=0}^n\prod_{k=0}^{T_i}\left(\dfrac{(c_{ik}{}^L)^{1-s}(c_{ik}{}^U)^s}{\delta_{ik}}\right)^{\delta_{ik}}\left(\sum_{k=1}^{T_i}\delta_{ik}\right)^{\left(\sum_{k=1}^{T_i}\delta_{ik}\right)} = d(\delta,s) \qquad (5.3.9)$$

i.e., $g_0(x,s) \geq d(\delta,s)$. (Proof)

Theorem 5.4(b)

If δ is a feasible vector for the dual programming (DP) problem, then $d(\delta,1) \geq d(\delta,0)$.

Proof:

We have $c_{ik}{}^U \geq c_{ik}{}^L$, for all k, (k=1,2,.......,T_0).

Or $\quad (c_{ik}{}^L)^{1-1}(c_{ik}{}^U)^1 \geq (c_{ik}{}^L)^{1-0}(c_{ik}{}^U)^0$

Or $\quad \dfrac{(c_{ik}{}^L)^{1-1}ik^U)^1}{\delta_{ik}} \geq \dfrac{(c_{ik}{}^L)^{1-0}(c_{ik}{}^U)^0}{\delta_{ik}}$

Or $\quad \left(\dfrac{(c_{ik}{}^L)^{1-1}(c_{ik}{}^U)^1}{\delta_{ik}}\right)^{\delta_{ik}} \geq \left(\dfrac{(c_{ik}{}^L)^{1-0}(c_{ik}{}^U)^0}{\delta_{ik}}\right)^{\delta_{ik}}$

Or $\quad \prod_{k=0}^{T_i} \left(\dfrac{(c_{ik}{}^L)^{1-1}(c_{ik}{}^U)^1}{\delta_{ik}}\right)^{\delta_{ik}} \geq \prod_{k=0}^{T_0} \left(\dfrac{(c_{ik}{}^L)^{1-0}(c_{ik}{}^U)^0}{\delta_{ik}}\right)^{\delta_{ik}}$

Or $\quad \prod_{i=1}^{n} \prod_{k=0}^{T_i} \left(\dfrac{(c_{ik}{}^L)^{1-1}(c_{ik}{}^U)^1}{\delta_{ik}}\right)^{\delta_{ik}} \left(\sum_{k=1}^{T_i} \delta_{ik}\right)^{(\sum_{k=1}^{T_i} \delta_{ik})}$

$$\geq \prod_{i=1}^{n} \prod_{k=0}^{T_0} \left(\dfrac{(c_{ik}{}^L)^{1-0}(c_{ik}{}^U)^0}{\delta_{ik}}\right)^{\delta_{ik}} \left(\sum_{k=1}^{T_i} \delta_{ik}\right)^{(\sum_{k=1}^{T_i} \delta_{ik})}$$

i.e., $\quad d(\delta,1) \geq d(\delta,0)$. (Proof) $\hfill (5.3.10)$

5.3.1 Solution procedure of the fuzzy model by geometric programming (GP) technique:

When $\tilde{a} = (a_1, a_2, a_3)$, $\tilde{H} = (H_1, H_2, H_3)$, $\tilde{\theta} = (\theta_1, \theta_2, \theta_3)$ and $\tilde{W} = (W_1, W_2, W_3)$ are triangular fuzzy number (TFN), then the fuzzy model is

$$\widetilde{\text{Min}} \quad \text{TAC(D,S,q)} = \frac{SD}{q} + \frac{\tilde{a}\tilde{H}q^2}{6D} + \tilde{\theta}D^{1-x}S^{-1} \qquad (5.3.1.1)$$

Subject to $\quad w_0 q \lesssim \tilde{W}$,

$$D, S, q > 0.$$

Using nearest interval approximation (NIA) method, the interval number corresponding triangular fuzzy number (TFN) $\tilde{a} = (a_1, a_2, a_3)$ is $[\frac{a_1+a_2}{2}, \frac{a_3+a_2}{2}] = [a_L, a_U]$. Similarly interval number corresponding $\tilde{H}, \tilde{\theta}$ and $\tilde{W}$ are $[\frac{H_1+H_2}{2}, \frac{H_3+H_2}{2}] = [H_L, H_U]$, $[\frac{\theta_1+\theta_2}{2}, \frac{\theta_3+\theta_2}{2}] = [\theta_L, \theta_U]$ and $[\frac{W_1+W_2}{2}, \frac{W_3+W_2}{2}] = [W_L, W_U]$ respectively. The problem (5.3.1.1) reduces to

Min $\qquad$ TAC(D,S,q) $= \frac{SD}{q} + \frac{[a_L, a_U][H_L, H_U]q^2}{6D} + [\theta_L, \theta_U]D^{1-x}S^{-1}$ $\qquad$ (5.3.1.2)

Subject to $\quad w_0 q \leq [W_L, W_U]$,

$\qquad$ D, S, q > 0.

Which is equivalent to

Min $\qquad$ TAC(D,S,q) $= \frac{SD}{q} + \frac{\hat{a}\hat{H}q^2}{6D} + \hat{\theta}D^{1-x}S^{-1}$ $\qquad$ (5.3.1.3)

Subject to $\quad w_0 q \lesssim \hat{W}$,

$\qquad$ D, S, q > 0.

Where $\hat{a} \in [a_L, a_U], \hat{H} \in [H_L, H_U], \hat{\theta} \in [\theta_L, \theta_U]$ and $\hat{W} \in [W_L, W_U]$.

According to nearest interval approximation (NIA) method and parametric interval-valued function, the fuzzy model (5.3.1.3) reduces to a parametric geometric programming by replacing $\hat{a} = a_L^{1-s}a_U^{s}$, $\hat{H} = H_L^{1-s}H_U^{s}$, $\hat{\theta} = \theta_L^{1-s}\theta_U^{s}$ and $\hat{W} = W_L^{1-s}W_U^{s}$ where s $\in$ [0, 1] as follows;

Min $\qquad$ TAC(D,S,q) $= \frac{SD}{q} + \frac{(a_L^{1-s}a_U^{s})(H_L^{1-s}H_U^{s})q^2}{6D} + (\theta_L^{1-s}\theta_U^{s})D^{1-x}S^{-1}$ $\qquad$ (5.3.1.4)

Subject to $\quad w_0 q \lesssim (W_L^{1-s}W_U^{s})$,

$\qquad$ D,S,q > 0.

Corresponding dual problem of (5.3.1.4) is given by

Max d(δ, s) $= (\frac{1}{\delta_1})^{\delta_1}(\frac{(a_L^{1-s}a_U^{s})(H_L^{1-s}H_U^{s})}{6\delta_2})^{\delta_2}(\frac{(\theta_L^{1-s}\theta_U^{s})}{\delta_3})^{\delta_3}(\frac{w_0}{(W_L^{1-s}W_U^{s})\delta_{01}})^{\delta_{01}}\delta_{01}^{\delta_{01}}$

Subject to

$\qquad \delta_1 + \delta_2 + \delta_3 = 1$ $\qquad$ (5.3.1.5)

$\qquad \delta_1 - \delta_3 = 0$

$\qquad \delta_1 - \delta + (1-x)\delta_3 = 0$

$\qquad -\delta_1 + 2\delta_2 + \delta_{01} = 0$

$\delta_1, \delta_2, \delta_3, \delta_{01} \geq 0.$

From (5.3.1.5) we get $\delta_1 = \frac{1}{4-x}, \delta_2 = \frac{2-x}{4-x}, \delta_3 = \frac{1}{4-x}$, and $\delta_{01} = \frac{2x-3}{4-x}$.

Putting the values in the objective function (5.3.1.5) we get the optimal solution of dual problem. The values of D, S and q are obtained by using the primal dual relation as follows:

From primal dual relation we get

$$\frac{SD}{q} = \delta_1^{\ *} \times d^*(\delta),$$

$$\frac{(a_L{}^{1-s}a_U{}^s)(H_L{}^{1-s}H_U{}^s)q^2}{6D} = \delta_2^{\ *} \times d^*(\delta),$$

$$(\theta_L{}^{1-s}\theta_U{}^s)D^{1-x}S^{-1} = \delta_3^{\ *} \times d^*(\delta),$$

$$\frac{w_0 q}{(W_L{}^{1-s}W_U{}^s)} = 1.$$

The optimal solutions of the model through the parametric approach is given by

$$S^* = \frac{6\omega_1{}^*\omega_2{}^*d^*(\delta,s)^2}{(a_L{}^{1-s}a_U{}^s)(H_L{}^{1-s}H_U{}^s)},$$

$$D^* = \frac{(a_L{}^{1-s}a_U{}^s)(H_L{}^{1-s}H_U{}^s)q^2}{6\omega_2{}^*d^*(\delta,s)},$$

$$q^* = \frac{(W_L{}^{1-s}W_U{}^s)}{w_0}.$$

Where

$$d^*(\delta,s) = (4-x)^{\frac{1}{4-x}} \left(\frac{(a_L{}^{1-s}a_U{}^s)(H_L{}^{1-s}H_U{}^s)(4-x)}{(2-x)6}\right)^{\frac{2-x}{4-x}} \left((\theta_L{}^{1-s}\theta_U{}^s)(4-x)\right)^{\frac{1}{4-x}} \left(\frac{w_0(4-x)}{(W_L{}^{1-s}W_U{}^s)(2x-3)}\right)^{\frac{2x-3}{4-x}} \times$$

$$(\frac{2x-3}{4-x})^{\frac{2x-3}{4-x}}.$$

5.4 Numerical Example and Solution

A manufacturing company produces a machine. It is given that the inventory carrying cost of the machine is \$15 per unit per year. The production cost of the machine varies inversely with the demand and set-up cost. From the past experience, the production cost of the machine is $120D^{-3}S^{-1}$ where D is the demand rate and S is set-up cost. Storage space area per unit time

(w_0) and total storage space area (W) are 100 sq. ft. and 2000 sq. ft. respectively. Determine the demand rate (D), set-up cost (S), production quantity (q), and optimum total average cost (TAC) of the production system.

Then the input value of the model (5.1.4.1) is

Table-5.1(Crisp input values)

a	H	x	θ	w_0	W
7	15	1.75	120	100	2000

Then the model is of the form

$$\text{Min} \quad \text{TAC(D,S,q)} = \frac{SD}{q} + \frac{105q^2}{6D} + 120D^{-0.75}S^{-1}$$

Subject to $\quad 100q \leq 2000, \quad D, S, q > 0.$ $\hspace{3cm}$ (5.4.1)

And the optimal solution of the crisp model, as follows

Table-5.2 (Optimal solution for crisp model)

Crisp model	S^*	D^*	q^*	$TAC^*(S^*,D^*,q^*)\$$
GP	0.684	4048	20	140.517
NLP	0.685	4047	20	140.685

When the input data of inventory model is taken as triangular fuzzy number (TFN) i.e.. $\tilde{a}$ = (5,7,9), $\tilde{H}$ = (13, 15, 17), $\tilde{\theta}$ = (116,120,124) and $\tilde{W}$ = (1800, 2000, 2200).

Using nearest interval approximation (NIA) method, we get the corresponding interval number and interval-valued function

i.e.,

$$\tilde{a} \approx [6, 8], \Rightarrow \hat{a} = (6)^{1-s}(8)^s \in [6, 8],$$

$$\tilde{H} \approx [14, 16], \Rightarrow \hat{H} = (14)^{1-s}(16)^s \in [14, 16],$$

$$\tilde{\theta} \approx [118,122], \Rightarrow \hat{\theta} = (118)^{1-s}(122)^s \in [118,122],$$

$\widetilde{W} \approx [1900, 2100], \Rightarrow \widehat{W} = (1900)^{1-s}(2100)^s \in [1900, 2100]$, where s $\in$ [0, 1].

Then the optimal solution of the fuzzy model by interval-valued parametric geometric programming (PGP) is presented in Table 5.3.

Table- 5.3(Optimal Solution for Fuzzy Inventory Model)

s	S^*	D^*	q^*	$TAC^*(S^*,D^*,q^*)\$$
0.0	0.820	2983.86	21.00	119.801
0.1	0.786	3175.16	20.79	123.060
0.2	0.753	3378.71	20.58	126.396
0.3	0.722	3595.32	20.38	129.924
0.4	0.693	3825.81	20.18	133.735
0.5	0.664	4071.08	19.97	137.533
0.6	0.637	4332.07	19.78	141.519
0.7	0.610	4609.80	19.58	145.473
0.8	0.585	4905.33	19.38	149.793
0.9	0.561	5219.81	19.19	154.196
1.0	0.538	5554.45	19.00	158.767

For s = 0, the lower bound of the interval value of the parameter is used to find the optimal solution. For s = 1, the upper bound of interval value of the parameter is used for the optimal solution. These results yield the lower and upper bounds of the optimal solution. The main advantage of the proposed technique is that one can get the intermediate optimal result using proper value s.

Here we have given a rough sketch, which shown how change the value of $TAC^*(S^*, D^*, q^*)$ for difference values of s.

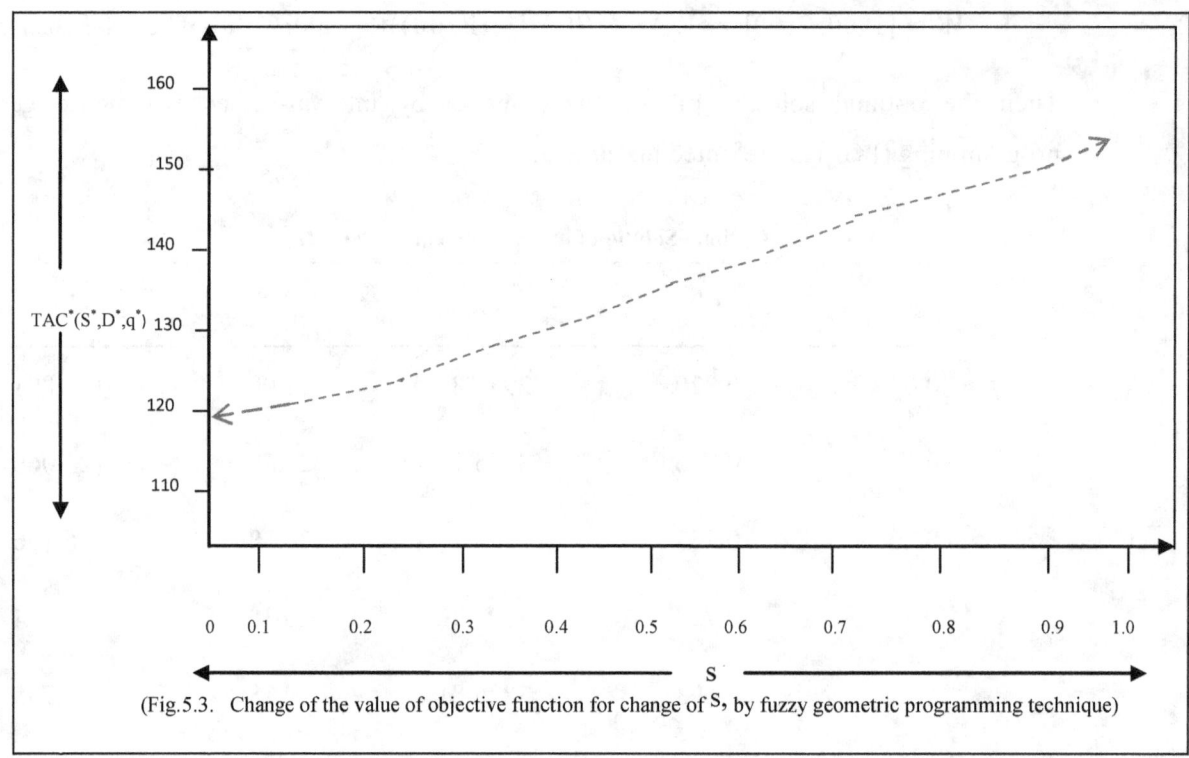

(Fig.5.3. Change of the value of objective function for change of s, by fuzzy geometric programming technique)

5.5   Sensitivity Analysis

Effects, for increment the parameter "s".

(1) For increasing value of "s", set-up cost S^* is decreasing.

(2) For increasing value of "s", demand rate D^* is increasing.

(3) For increasing value of "s , Production quantity q^* is decreasing.

(4) For increasing value of "s", Total average cost $TAC^*(S^*, D^*, q^*)$ is increasing.

5.6   Conclusion

In this chapter, we have proposed a real life inventory problem in fuzzy environment and solved by parametric geometric programming (PGP) technique, also presented a numerical example along with sensitivity analysis. The inventory model is developed with unit production cost, time depended holding cost, without shortages. This model has been developed for a single item. Here we have considered the coefficients are triangular fuzzy number (TFN). In future, the other type of fuzzy numbers such as piecewise linear hyperbolic, L-R fuzzy number, trapezoidal fuzzy number (TrFN), parabolic flat fuzzy number (PfFN), pentagonal fuzzy number (PFN), hexagonal fuzzy umber (HFN), generalized fuzzy number (GFN) etc. can be considered for the coefficients and then model can be easily solved. The model is developed under negligible lead time. But lead time plays very important roles in an inventory. So in future, consideration lead time in an inventory the model would be more interesting and challenging.

Chapter 6

A Fuzzy EOQ Model with Unit Production Cost, Time Depended Holding Cost, Without Shortages under a Space Constraints: A Fuzzy Geometric Programming (FGP) Approach

In most of the existing literature, inventory related costs are assumed in deterministic nature. But in real situation costs are usually imprecise in nature due to the influence of various uncontrollable factors. So inventory production cost may also depend on some parameters like demand and set-up cost. The study of inventory model where demand rates varies with time is the last century.

In this chapter, an economic order quantity (EOQ) model with unit production cost, time dependent holding cost, without shortages is formulated and solved. *In most real world situation, the objective and constraint function of the model are imprecise in nature, so the coefficients are imposed here in fuzzy environment. The problem is solved by using fuzzy max-min geometric programming (GP) technique and fuzzy parametric geometric programming (GP) technique respectively. Here a numerical example is presented to illustrate the model. Sensitivity analysis is also presented here.*

6.1   Mathematical Model

An EOQ model is developed under the following notations and assumptions.

6.1.1   Notations

I(t):Inventory level at any time, t ≥ 0.

D: Constant demand per unit time.

T: Cycle of length.

S: Set-up cost per batch.

H(t): Time dependent holding cost per unit per unit time.

q: Production quantity per batch.

f(D,S): Unit production cost per cycle.

TAC(D,S,q):Total average cost per unit time

w_0: Space area per unit quantity.

W: Total storage space area.

6.1.2 Assumptions

a) The inventory system involves only one item.

b) The replenishment occurs instantaneously at infinite rate.

c) The lead time is negligible.

d) Demand rate is constant.

e) Inflation is negligible.

f) The model is developed with no shortages.

g) The unit production cost is continuous function of demand and Set-up cost and take

the form: $f(D,S) = \theta D^{-x} S^{-1}$, θ, $x \in \mathbb{R}$ (>0).

h) Holding cost is time dependent, $H(t) = a.\,Ht$, a, $H \in \mathbb{R}$ (>0).

6.1.3 Crisp model

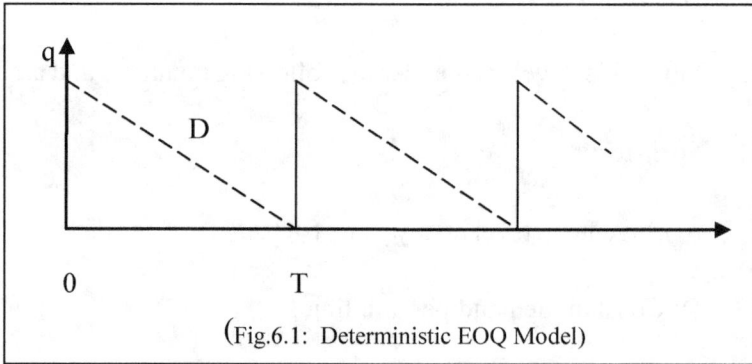

(Fig.6.1: Deterministic EOQ Model)

The differential equations describing the above model as follows,

$$\frac{dI(t)}{dt} = -D, \qquad (0 \leq t \leq T) \qquad\qquad (6.1.3.1)$$

With the boundary condition I(0) = q, I(T) = 0.

The solution of (6.1.3.1) is obtained as

$$I(t) = q - Dt. \tag{6.1.3.2}$$

Also there are

$$T = q/D.$$

Now inventory holding cost $= H \int_0^T at . I(t) dt = \frac{aHq^3}{6D^2}. \tag{6.1.3.3}$

Total inventory related cost per cycle = set-up cost + holding cost + production cost

$$= S + \frac{aHq^3}{6D^2} + f(D,S). \tag{6.1.3.4}$$

So total average cost per cycle is given by

$$TAC(D,S,q) = \frac{SD}{q} + \frac{aHq^3}{6D^2} + \theta D^{1-x} S^{-1}. \tag{6.1.3.5}$$

And storage area $= w_0 q$

Hench the inventory model can be written as

$$\text{Min} \qquad TAC(D,S,q) = \frac{SD}{q} + \frac{aHq^2}{6D} + \theta D^{1-x} S^{-1} \tag{6.1.3.6}$$

Subject to $\quad w_0 q \le W, \qquad D, S, q > 0.$

6.1.4 Fuzzy model

When the coefficients become fuzzy parameters, the crisp model (6.1.3.6) written to be a fuzzy model, as following form

$$\widetilde{Min} \qquad TAC(D,S,q) = \frac{SD}{q} + \frac{\tilde{a}\tilde{H}q^2}{6D} + \tilde{\theta} D^{1-x} S^{-1}$$

Subject to $\widetilde{w_0} q \lesssim W, \qquad D,S,q > 0. \tag{6.1.4.1}$

6.2 Mathematical Analysis

Consider a non-linear programming as follows,

(P)   Min $g_0(x)$

        Subject to $g_i(x) \le 1 \quad (1 \le i \le n), \tag{6.2.1}$

$x > 0$.

Its objective and constraints of the form

$$g_i(x) = \sum_{k=1}^{T_i} C_{ik} \prod_{j=1}^{m} x_j{}^{\alpha_{ikj}} \quad (0 \leq i \leq n)$$

$$x_j > 0, \quad (j = 1, 2, \ldots, m)$$

Here C_{ik} (> 0), ($k = 1, 2, \ldots, T_0$) and α_{ikj} be any real numbers.

When the objective and constraint goals, coefficients and exponents become fuzzy sets and fuzzy numbers respectively, then we transform (P) into a fuzzy geometric programming as follows,

(P) $\quad \widetilde{Min} \quad g_0(x)$

Subject to $\ g_i(x) \lesssim 1 \quad (1 \leq i \leq n)$ $\qquad\qquad\qquad\qquad\qquad$ (6.2.2)

$$x > 0,$$

Its objective and constraints of the form $g_i(x) = \sum_{k=1}^{T_i} \tilde{c}_{ik} \prod_{j=1}^{m} x_j{}^{\tilde{\alpha}_{ikj}}$ ($0 \leq i \leq n$), are all posynomials of x in which coefficients $\tilde{c}_{ik}$ and indexes $\tilde{\alpha}_{ikj}$ are fuzzy numbers.

6.2.1 Some definition and theorem

Definition 6.2.1.1 For nth parabolic flat fuzzy number $(a_1, a_2, a_3, a_4)_{PfFN}$ containing the coefficients $\tilde{c}_{ik}$ ($0 \leq i \leq n$; $1 \leq k \leq T_i$), the membership function of $\tilde{c}_{ik}$ is

$$\mu_{\tilde{c}_{ik}}(\tilde{c}_{ik}) = \begin{cases} 1 - \left(\frac{a_2 - c_{ik}}{a_2 - a_1}\right)^n & for\, a_1 \leq c_{ik} \leq a_2 \\ 1 & for\, a_1 \leq c_{ik} \leq a_2 \\ 1 - \left(\frac{c_{ik} - a_3}{a_4 - a_3}\right)^n & for\, a_3 \leq c_{ik} \leq a_4 \\ 0 & for\ otherwise. \end{cases} \qquad (6.2.1.1)$$

Similarly, we can determine the membership function of the indexes $\tilde{\alpha}_{ikj}$ ($0 \leq i \leq n$; $1 \leq k \leq T_i$; $1 \leq j \leq m$).

Note:

(a)   when n=1, $\tilde{c}_{ik}$ become Trapezodial Fuzzy Number (TrFN),

(b)   when n=1, and $a_3 = a_4$, $\tilde{c}_{ik}$ become Triangular Fuzzy Number (TFN),

(c)   when n=2, $\tilde{c}_{ik}$ become Parabolic flat Fuzzy Number (PfFN),

(d)   when n=2, and $a_3=a_4$, $\tilde{c}_{ik}$ become Parabolic Fuzzy Number (PrFN).

Definition 6.2.1.2 Here δ-cut of $\tilde{c}_{ik}$ $(0 \leq i \leq n; 1 \leq k \leq T_i)$ is given by

$$\mu_{\tilde{c}_{ik}}{}^{-1}(\delta) = [\,\mu_{\tilde{c}_{ikL}}{}^{-1}(\delta), \mu_{\tilde{c}_{ikR}}{}^{-1}(\delta)\,] = [a_1 + \sqrt[n]{1-\delta}(a_2 - a_1), a_4 - \sqrt[n]{1-\delta}(a_4 - a_3)]. \qquad (6.2.1.2)$$

Similarly, we can determine the δ-cut of $\tilde{\alpha}_{ikj}$ $(0 \leq i \leq n; 1 \leq k \leq T_i; 1 \leq j \leq m)$.

Proposition 6.2.1.3 When the coefficient and indexes of the fuzzy geometric programming problem are taken as fuzzy numbers

$$\widetilde{Min} \qquad \sum_{k=1}^{T_i} \tilde{c}_{ok} \prod_{j=1}^{m} x_j{}^{\tilde{\alpha}_{okj}}$$

Subject to $\sum_{k=1}^{T_i} \tilde{c}_{ik} \prod_{j=1}^{m} x_j{}^{\tilde{\alpha}_{ikj}} \lesssim 1$ $(1 \leq i \leq n)$, $(6.2.1.3)$

         $x_j > 0.$

   Using δ-cut of fuzzy numbers coefficients and indexes, the above problem is reduces to

$$\widetilde{Min} \qquad \sum_{k=1}^{T_0} [\,\mu_{\tilde{c}_{okL}}{}^{-1}(\delta), \mu_{\tilde{c}_{okR}}{}^{-1}(\delta)\,] \prod_{j=1}^{m} x_j{}^{[\mu_{\tilde{\alpha}_{okjL}}{}^{-1}(\delta), \mu_{\tilde{\alpha}_{okjR}}{}^{-1}(\delta)]}$$

Subject to $\sum_{k=1}^{T_i} [\,\mu_{\tilde{c}_{ikL}}{}^{-1}(\delta), \mu_{\tilde{c}_{ikR}}{}^{-1}(\delta)\,] \prod_{j=1}^{m} x_j{}^{[\mu_{\tilde{\alpha}_{ikjL}}{}^{-1}(\delta), \mu_{\tilde{\alpha}_{ikjR}}{}^{-1}(\delta)]} \lesssim 1$ $(1 \leq i \leq n)$,

     $x_j > 0.$

Which is equivalent to

$$\widetilde{Min} \sum_{k=1}^{T_0} \mu_{\tilde{c}_{okL}}{}^{-1}(\delta) \prod_{j=1}^{m} x_j{}^{\mu_{\tilde{\alpha}_{okjS}}{}^{-1}}(\delta)$$

subject to $\sum_{k=1}^{T_i} \mu_{\tilde{c}_{ikL}}{}^{-1}(\delta) \prod_{j=1}^{m} x_j{}^{\mu_{\tilde{\alpha}_{ikjS}}{}^{-1}}(\delta) \leq 1.$ $(1 \leq i \leq n)$ $(6.2.1.4)$

Where

$$\mu_{\tilde{\alpha}_{ikjS}}{}^{-1}(\delta) = \begin{cases} \mu_{\tilde{\alpha}_{ikjL}}{}^{-1}(\delta) & when\, \tilde{\alpha}_{ikjL} > 0, \\ \mu_{\tilde{\alpha}_{ikjR}}{}^{-1}(\delta) & when\, \tilde{\alpha}_{ikjL} < 0. \end{cases} \qquad (1 \leq i \leq n)$$

Definition 6.2.1.4 For any $x \in \mathbb{R}^m$ and feasible index $d_i \in \mathbb{R}$ ($\mathbb{R}$ is the real number set), if $g_i(x, \delta) = \sum_{k=1}^{T_i} \mu_{\tilde{c}_{ikL}}{}^{-1}(\delta) \prod_{j=1}^{m} x_j{}^{\mu_{\tilde{\alpha}_{ikjS}}{}^{-1}}(\delta) \leq 1 (1 \leq i \leq n)$, then the linear membership function are given by

$$\mu_0(g_0(x, \delta)) = \begin{cases} 1 & if\, g_0(x, \delta) \leq z_0, \\ \left(\frac{z_0 + d_0 - g_0(x, \delta)}{d_0}\right) & if\, z_0 \leq g_0(x, \delta) \leq z_0 + d_0, \\ 0 & if\, g_0(x, \delta) \geq z_0 + d_0, \end{cases} \qquad (6.2.1.5)$$

$$\mu_i(g_i(x,\delta)) = \begin{cases} 1 & if\, g_i(x,\delta) \le z_0, \\ \left(\frac{1+d_i-i(x,\delta)}{d_0}\right) & if\ 1 \le g_i(x,\delta)\ \le 1+d_i, \\ 0 & if\, g_i(x,\delta) \ge 1+d_i, \end{cases} \qquad (6.2.1.6)$$

Based on Zimmerman, first finding δ-cut of the fuzzy numbers in coefficients and indexes then we built membership functions of both objective and constraints goals and using max-min operator the problem (6.2.1.4) reduced to a fuzzy non-linear programming(FNLP) problem

Max $\qquad \lambda$

Subject to $\quad \mu_i(\sum_{k=1}^{T_i} \mu_{\tilde{c}_{ikL}}^{-1}(\delta) \prod_{j=1}^{m} x_j^{\mu_{\tilde{\alpha}_{ikjS}}^{-1}}(\delta)) \ge \lambda \qquad (1 \le i \le n),$ $\qquad (6.2.1.7)$

$\qquad x > 0, \quad \lambda, \delta \in [0,1].$

Which is equivalent to a geometric programming problem, with parameters λ, δ variation

Min $\qquad \lambda^{-1}$

Subject to $\quad \mu_i(\sum_{k=1}^{T_i} \mu_{\tilde{c}_{ikL}}^{-1}(\delta) \prod_{j=1}^{m} x_j^{\mu_{\tilde{\alpha}_{ikjS}}^{-1}}(\delta)) \ge \lambda \quad (1 \le i \le n),$ $\qquad (6.2.1.8)$

$\qquad x > 0, \quad \lambda, \delta \in [0,1],$

Theorem 6.2.1.5 Let the membership function $\mu_i(g_i(x,\delta))$, $\mu_{\tilde{c}_{ik}}(c_{ik})$, $\mu_{\tilde{\alpha}_{ikj}}(\alpha_{ikj})$ be all continuous and strictly monotone. Then (6.2.1.8) is equivalent with

Min $\qquad \lambda^{-1}$

Subject to $\quad \dfrac{\sum_{k=1}^{T_i} \mu_{\tilde{c}_{ikL}}^{-1}(\delta) \prod_{j=1}^{m} x_j^{\mu_{\tilde{\alpha}_{ikjS}}^{-1}}(\delta)}{\mu_i^{-1}(\delta)} \le 1,$

$\qquad x > 0, \quad \lambda, \delta \in [0,1], \qquad (0 \le i \le n, 1 \le j \le m).$

Proof: Pls. see the reference S. Islam, T.K. Roy (2006).

Corollary 6.2.1.6 Let the membership function $\mu_i(g_i(x,\delta))$, $\mu_{\tilde{c}_{ik}}(c_{ik})$, $\mu_{\tilde{\alpha}_{ikj}}(\alpha_{ikj})$ be all continuous and strictly monotone and the problem is

Min $\qquad \lambda^{-1}$

Subject to $\quad \dfrac{\sum_{k=1}^{T_i} \mu_{\tilde{c}_{ikL}}^{-1}(\delta) \prod_{j=1}^{m} x_j^{\mu_{\tilde{\alpha}_{ikjS}}^{-1}}(\delta)}{\mu_i^{-1}(\delta)} \le 1,$ $\qquad (6.2.1.9)$

$$x > 0, \quad \lambda, \delta \in [0,1], \quad (0 \le i \le n, \ 1 \le j \le m).$$

Which is a classical posynomial geometric programming with parameters γ and δ.

Its dual form is

$$\text{Max} \qquad d(\omega) = \left(\frac{\lambda^{-1}}{\omega_{00}}\right)^{\omega_{00}} \prod_{i=0}^{n} \prod_{k=1}^{T_i} \left(\frac{\mu_{\tilde{c}_{ik}}^{-1}(\delta)/\mu_i^{-1}(\lambda)}{\omega_{ik}}\right)^{\omega_{ik}} \qquad (6.2.1.10)$$

Subject to

$$\omega_{00} = 1,$$

$$\omega_{00} = \sum_{k=1}^{T_0} \omega_{0k}$$

$$(\Gamma(\delta))^T \omega = 0, \qquad \lambda, \delta \in [0,1],$$

$$\omega \ge 0.$$

Where $\omega_{ik} = \omega_{ik}(\delta, \lambda)$,

and
$$\Gamma(\delta) = \begin{pmatrix} \tilde{\alpha}_{011}^{-1}(\delta) \cdots & \tilde{\alpha}_{01l}^{-1}(\delta) \cdots & \tilde{\alpha}_{01m}^{-1}(\delta) \\ \cdots & \cdots & \cdots \\ \tilde{\alpha}_{0J_01}^{-1}(\delta) \cdots & \tilde{\alpha}_{0J_01}^{-1}(\delta) \cdots & \tilde{\alpha}_{0J_01}^{-1}(\delta) \\ \cdots \cdots \cdots & & \\ \tilde{\alpha}_{p11}^{-1}(\delta) \cdots & \tilde{\alpha}_{p1l}^{-1}(\delta) \cdots & \tilde{\alpha}_{p1m}^{-1}(\delta) \\ \cdots & \cdots & \cdots \\ \tilde{\alpha}_{pJ_p1}^{-1}(\delta) \cdots & \tilde{\alpha}_{pJ_p1}^{-1}(\delta) \cdots & \tilde{\alpha}_{pJ_p1}^{-1}(\delta) \end{pmatrix}.$$

6.3   Solution Procedure

6.3.1 Fuzzy max-min geometric programming (GP) technique on for solving fuzzy model

When coefficient and exponents are taken as a triangular fuzzy number i.e., in general $\tilde{\mathcal{L}} = (a_1, a_2, a_3)$, then the δ-cut of the fuzzy number $\tilde{\mathcal{L}}$, is given by $\mathcal{L}(\delta) = [a_1 + \delta(a_2 - a_1), \ a_3 - \delta(a_3 - a_2)]$, $\delta \in [0,1]$.

Taking the membership function as in (6.2.1.5) and (6.2.1.6) the problem (6.1.4.1) reduced to (by using corollary (6.2.1.6)),

Min λ^{-1}

Subject to $\dfrac{-SDq^{-1}-\left(H^1+\delta(H^2-H^1)\right)aq^2D^{-1}/6-(\theta^1+\delta(\theta^2-\theta^1))D^{1-x}S^{-1}}{(-(z_0+d_0-1)+d_0\lambda)} \leq 1$ (6.3.1.1)

$\dfrac{(w_0{}^1+\delta(w_0{}^2-w_0{}^1))q}{W+d_1\lambda} \leq 1$

$D, S, q > 0, \qquad \gamma, \delta \in [0, 1].$

The dual problem of (6.3.1.1) is given by

$$\text{Max } d(\omega) = \left(\frac{\lambda^{-1}}{\omega_{00}}\right)^{\omega_{00}} \left(\frac{\frac{1}{\omega_{01}}}{(z_0+d_0-1)-d_0\lambda}\right)^{\omega_{01}} \left(\frac{\frac{\left(H^1+\delta(H^2-H^1)\right)a}{6\omega_{02}}}{(z_0+d_0-1)-d_0\lambda}\right)^{\omega_{02}} \left(\frac{\frac{(\theta^1+\delta(\theta^2-\theta^1))}{\omega_{03}}}{(z_0+d_0-1)-d_0\lambda}\right)^{\omega_{03}} \left(\frac{\frac{(w_0{}^1+\delta(w_0{}^2-w_0{}^1))}{\omega_{11}}}{W+d_1\lambda}\right)^{\omega_{11}}$$

Subject to

$\omega_{00} = 1,$

$\omega_{01}+\omega_{02}+\omega_{03}+\omega_{11} = \omega_{00},$ (6.3.1.2)

$\omega_{01} - x\omega_{03} = 0$

$\omega_{01} - \omega_{02} + (1-x)\omega_{03} = 0$

$-\omega_{01} + 2\omega_{02} + \omega_{11} = 0$

$\omega_{01}, \omega_{02}, \omega_{03}, \omega_{11} \geq 0.$

Solving the system of linear equations of (6.3.1.2) we get $\omega_{01} = \dfrac{1}{4-x}, \omega_{02} = \dfrac{2-x}{4-x}, \omega_{03} = \dfrac{1}{4-x}, \omega_{11} = \dfrac{2x-3}{4-x}.$

Putting the values in the objective function of the problem (6.3.1.2), we get

$$d(\omega) = \lambda^{-1}\left(\frac{4-x}{(z_0+d_0-1)-d_0\lambda}\right)^{\frac{1}{4-x}} \left(\frac{\frac{\left(H^1+\delta(H^2-H^1)\right)a(4-x)}{6(2-x)}}{(z_0+d_0-1)-d_0\lambda}\right)^{\frac{2-x}{4-x}} \left(\frac{(\theta^1+\delta(\theta^2-\theta^1))(4-x)}{(z_0+d_0-1)-d_0\lambda}\right)^{\frac{1}{4-x}} \times$$

$$\left(\frac{(w_0{}^1+\delta(w_0{}^2-w_0{}^1))(4-x)/(2x-3)}{W+d_1\lambda}\right)^{\frac{2x-3}{4-x}} \left(\frac{2x-3}{4-x}\right)^{\left(\frac{2x-3}{4-x}\right)}.$$

We can obtained λ by the aid of $d(\omega) = \lambda^{-1}$. Then the above equation is reduces to

$$\left(\frac{4-x}{(z_0+d_0-1)-d_0\lambda}\right)^{\frac{1}{4-x}}\left(\frac{\frac{(H^1+\delta(H^2-H^1))a(4-x)}{6(2-x)}}{(z_0+d_0-1)-d_0\lambda}\right)^{\frac{2-x}{4-x}}\left(\frac{(\theta^1+\delta(\theta^2-\theta^1)(4-x)}{(z_0+d_0-1)-d_0\lambda}\right)^{\frac{1}{4-x}}\left(\frac{(w_0^1+\delta(w_0^2-w_0^1)(4-x)/(2x-3)}{W+d_1\lambda}\right)^{\frac{2x-3}{4-x}}$$

$$\left(\frac{2x-3}{4-x}\right)^{\frac{2x-3}{4-x}}=1. \tag{6.3.1.4}$$

Solving the above non-linear equation, for given $\delta \in [0, 1]$ by Newton-Raphson method, we obtain the value of λ^*. Putting the value of λ^*, we obtained the dual objective function.

Again from the relation between primal-dual variables, we get

$$\frac{-SDq^{-1}}{(-(z_0+d_0-1)+d_0\lambda^*)}=\frac{\omega_{01}{}^*}{\omega_{00}{}^*}=\frac{1}{4-x},$$

$$\frac{-(H^1+\delta(H^2-H^1))aq^2D^{-1}/6}{(-(z_0+d_0-1)+d_0\lambda^*)}=\frac{\omega_{02}{}^*}{\omega_{00}{}^*}=\frac{2-x}{4-x},$$

$$\frac{-(\theta^1+\delta(\theta^2-\theta^1))D^{1-x}S^{-1}}{(-(z_0+d_0-1)+d_0\lambda^*)}=\frac{\omega_{03}{}^*}{\omega_{00}{}^*}=\frac{1}{4-x},$$

$$\frac{(w_0^1+\delta(w_0^2-w_0^1))q}{W+d_1\lambda^*}=\frac{\omega_{11}{}^*}{\omega_{11}{}^*}=1. \tag{6.3.1.4}$$

Solving the above relation, we get

$$S^*=\frac{6\omega_1{}^*\omega_2{}^*((z_0+d_0-1)-d_0\lambda^*)^2}{a(H^1+\delta(H^2-H^1))}$$

$$D^*=\frac{a(H^1+\delta(H^2-H^1))q^2}{6\omega_2{}^*((z_0+d_0-1)-d_0\lambda^*)}$$

$$q^*=\frac{W+d_1\lambda^*}{(w_0^1+\delta(w_0^2-w_0^1))}.$$

6.3.2 Fuzzy parametric geometric programming (PGP) technique for solving fuzzy model:

Taking $\tilde{H}=H^1+\delta(H^2-H^1), \tilde{\theta}=\theta^1+\delta(\theta^2-\theta^1), \widetilde{w_0}=w_0^1+\delta(w_0^2-w_0^1)$ and $\tilde{W}=W^1+\delta(W^2-W^1)$ where $\alpha \in [0, 1]$ in (6.1.4.1). The model takes the reduces form as follows

$$\text{Min} \quad \text{TAC(D,S,q)}=\frac{SD}{q}+\frac{a(H^1+\delta(H^2-H^1))q^2}{6D}+(\theta^1+\delta(\theta^2-\theta^1))D^{1-x}S^{-1} \tag{6.3.2.1}$$

Subject to $(w_0^1+\delta(w_0^2-w_0^1))\,q \le (W^1+\delta(W^2-W^1))$, D, S, q > 0.

Applying GP technique the dual programming of the problem (6.3.2.1) is

$$\text{Max } d(\omega) = \left(\frac{1}{\omega_1}\right)^{\omega_1}\left(\frac{a(H^1+\delta(H^2-H^1))}{6\omega_2}\right)^{\omega_2}\left(\frac{\theta^1+\delta(\theta^2-\theta^1)}{\omega_3}\right)^{\omega_3}\left(\frac{w_0^1+\delta(w_0^2-w_0^1)}{(W^1+\delta(W^2-W^1))}\right)^{\omega_{01}}\omega_{01}^{\omega_{01}}$$

Subject to

$$\omega_1 + \omega_2 + \omega_3 = 1 \qquad\qquad (6.3.2.2)$$

$$\omega_1 - \omega_3 = 0$$

$$\omega_1 - \omega_2 + (1-x)\omega_3 = 0$$

$$-\omega_1 + 2\omega_2 + \omega_{01} = 0$$

$$\omega_1, \omega_2, \omega_3, \omega_{01} \geq 0.$$

Solving the above linear equations, we get $\omega_1 = \frac{1}{4-x}, \omega_2 = \frac{2-x}{4-x}, \omega_3 = \frac{1}{4-x}$ and $\omega_{01} = \frac{2x-3}{4-x}$.

Putting the values in the objective function of the problem (6.3.2.2), we get

$$d^*(\omega) = (4-x)^{\frac{1}{4-x}}\left(\frac{a\left(H^1+\delta(H^2-H^1)\right)(4-x)}{(2-x)6}\right)^{\frac{2-x}{4-x}}\left((\theta^1+\delta(\theta^2-\theta^1))(4-x)\right)^{\frac{1}{4-x}}.\times$$

$$\left(\frac{(w_0^1+\delta(w_0^2-w_0^1))(4-x)}{(W^1+\delta(W^2-W^1))(2x-3)}\right)^{\frac{2x-3}{4-x}}\left(\frac{2x-3}{4-x}\right)^{\frac{2x-3}{4-x}}$$

From primal dual relation, we have

$$\frac{SD}{q} = \omega_1^\star \times d^*(\omega),$$

$$\frac{a(H^1+\delta(H^2-H^1))q^2}{6D} = \omega_2^\star \times d^*(\omega),$$

$$(\theta^1+\delta(\theta^2-\theta^1))D^{1-x}S^{-1} = \omega_3^\star \times d^*(\omega),$$

$$\frac{(w_0^1+\delta(w_0^2-w_0^1))q}{(W^1+\delta(W^2-W^1))} = 1.$$

The optimal solution of the model through the parametric approach is given by

$$S^* = \frac{6\omega_1^\star\omega_2^\star d^*(\omega)^2}{a(H^1+\delta(H^2-H^1))},$$

$$D^* = \frac{a\left(H^1+\delta(H^2-H^1)\right)q^2}{6\omega_2^\star d^*(\omega)},$$

$$q^* = \frac{(W^1 + \delta(W^2 - W^1))}{(w_0{}^1 + \delta(w_0{}^2 - w_0{}^1))}. \tag{6.3.2.3}$$

6.4 Numerical Example and Solution

A manufacturing company produces a machine. It is given that the inventory carrying cost of the machine is \$15 per unit per year. The production cost of the machine varies inversely with the demand and set-up cost. From the past experience, the production cost of the machine is $120D^{-3}S^{-1}$ where D is the demand rate and S is set-up cost. Storage space area per unit time (w_0) and total storage space area (W) are 100 sq. ft. and 2000 sq. ft. respectively. Determine the demand rate (D), set-up cost (S), production quantity (q), and optimum total average cost (TAC) of the production system.

Then the crisp input values of the model (6.1.4.1) is

Table-6.1 (Crisp input data)

a	H	x	θ	w_0	W
7	15	1.75	120	100	2000

Then the model is of the form,

$$\text{Min} \qquad TAC(D,S,q) = \frac{SD}{q} + \frac{105q^2}{6D} + 120D^{-0.75}S^{-1}$$

Subject to $100q \leq 2000, \quad D, S, q > 0.$ (6.5.1)

And the optimal solution in crisp environment, as follows

Table-6.2(Optimal solution for crisp model)

Crisp model	S^*	D^*	q^*	$TAC^*(S^*,D^*,q^*)$
GP	0.684	4048	20	140.517
NLP	0.685	4047	20	140.685

For fuzzy max-min geometric programming technique, lets we consider $z_0 = 15.5$, fuzzy objective goal tolarence $d_0 = 1$ and total storage space area tolerance $d_1 = 100$, also taking H = (14, 16, 18), θ = (116,120,124), w_0 = (96,100,104), δ = 0.5, then from (6.3.1.10) we get λ = 0.007.

For fuzzy parametric geometric programming technique taking δ = 0.5, H = (14, 16, 18), θ = (116,120,124), w_0 = (96,100,104), and W = (2000, 2200, 2400).

And corresponding solution as follows,

Table-6.3(Optimal solution for fuzzy model)

Fuzzy model	S^*	D^*	q^*	$TAC^*(S^*, D^*, q^*)$
FGP(max-min)	0.677	4236	20.415	142.561
FGP(parametric)	0.662	4718	21.428	147.780

6.5   Sensitivity Analysis

6.5.1 We now examine to sensitivity analysis of the optimal solution of the model for changing "δ", keeping the other parameters is unchanged. The initial data are given from the above numerical example.

Table 6.4 (Sensitivity analysis)

Value of δ	% of change	FGP (max-min)	FGP (parametric)
0.1	-80	144.939	142.077
0.2	-60	144.295	142.983
0.3	-40	143.849	143.570
0.4	-20	143.108	144.349
0.5	0	142.561	144.830
0.6	+20	141.934	145.257
0.7	+40	141.484	144.542
0.8	+60	140.802	145.775
0.9	+80	140.271	145.989

Here we have given a rough sketch, which shown how change the value of $TAC^*(S^*, D^*, q^*)$ for different values of δ.

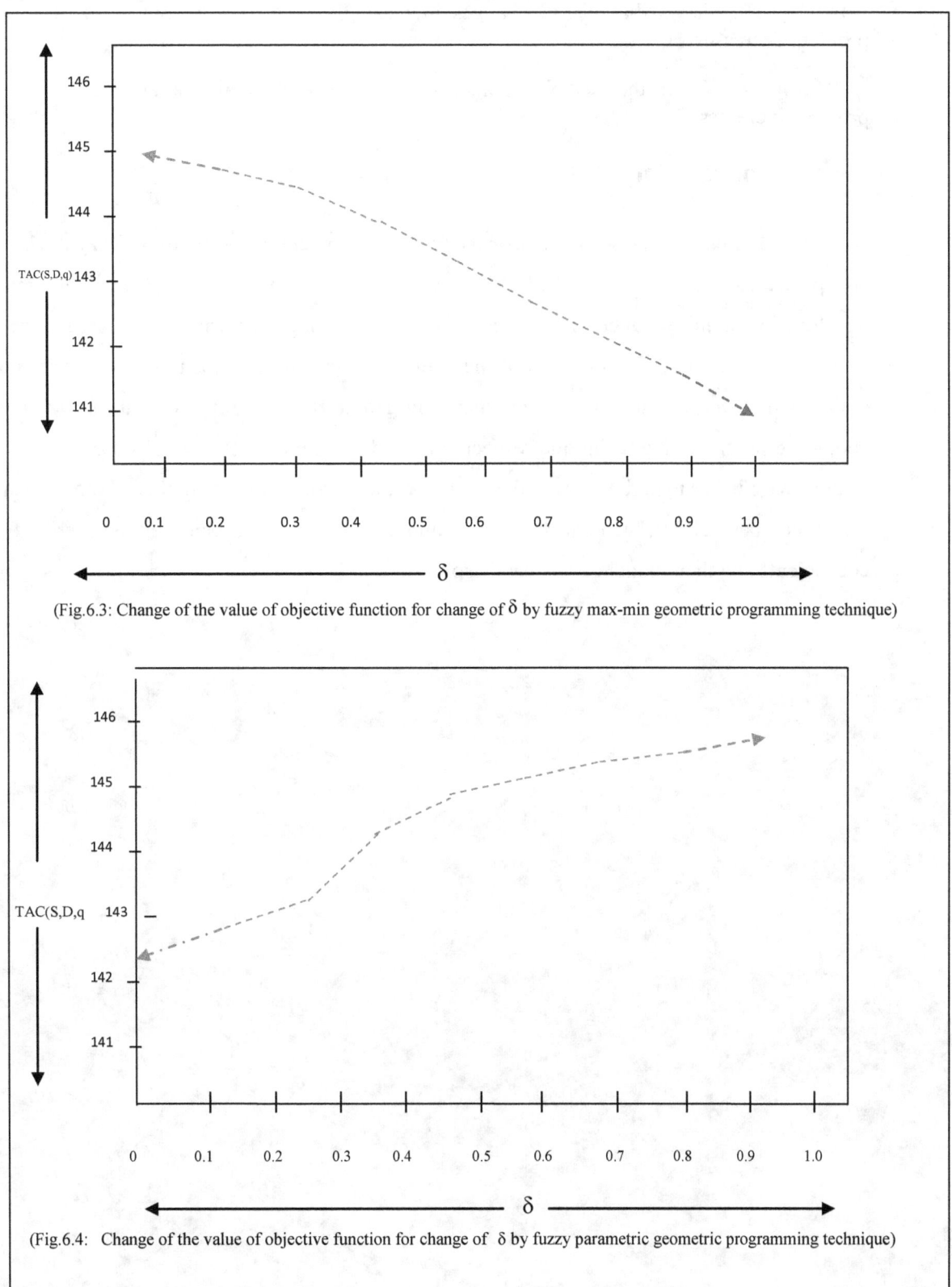

(Fig.6.3: Change of the value of objective function for change of δ by fuzzy max-min geometric programming technique)

(Fig.6.4: Change of the value of objective function for change of δ by fuzzy parametric geometric programming technique)

6.5.2   Effect, for increment parameters-:

1) Fig.6.3. Shows that if "δ" changes increasingly then total average cost of the given problem decreases.

2) Fig.6.4. Shows that if "δ" changes increasingly then total average cost of the given problem increases.

6.6   Conclusion

In this chapter, we have proposed a real life inventory problem in fuzzy environment and presented a numerical example along with sensitivity analysis. The inventory model is developed with unit production cost, time dependent holding cost, without shortages. Lead time is not considered in this model. Here the coefficients are taken as a triangular fuzzy number (TFN) and solved the model by fuzzy max-min geometric programming and fuzzy parametric geometric programming technique respectively. In future, the other type of fuzzy number, such as piecewise linear hyperbolic, L-R fuzzy number, trapezoidal fuzzy number (TrFN), hexagonal fuzzy number (HFN), generalized fuzzy number (GFN), etc. can be considered for the coefficients and then model can be easily solved.

Chapter 7

Fuzzy Inventory Model for Deteriorating Items, with Time Depended Demand with Shortages, and Fully Backlogging

In an inventory models deteriorations plays an important role. Deterioration is defined as decay or damage in the quality or quantity of the inventory. Foods, Drugs, pharmaceuticals etc. are deteriorating items. During inventory there have some losses of these deteriorating items, consequently this loss must be taken into account when analyzing the system. Shortages are also very important condition. There are several types of customers. At shortage period some customers are waiting for actual product and others do not it. In ordinary inventory model it considers all parameters like shortage cost, holding cost, deteriorating cost as a fixed. But in real life situation it will have some little fluctuations. So consideration of fuzzy in put values is more realistic.

In this chapter we have analyzed fuzzy inventory model for deterioration item with time dependent demand. *Shortages are allowed under fully backlogged. Fixed cost, deterioration cost, shortages cost, holding cost are the cost functions considered in this model. Fuzziness is applying by allowing the cost components. Here we have used triangular fuzzy number (TFN). One numerical solution of this model is obtained to verify optimal solution. Sensitivity analysis is also presented here. The purpose of the model is to minimize total average cost.*

7.1 Mathematical Model

An EOQ model is developed under the following notations and assumptions.

7.1.1 Notations

I(t):Inventory level at any time, t ≥ 0.

T: Cycle of length.

t_w: Time point, when demand rate start with $(c-dt)$..

t_1: Time point when stock level reaches to zero.

c_1: Fixed cost.

c_2: Shortages cost per unit per unit time.

c_3: Deteriorating cost per unit per unit time.

c_4: Holding cost per unit per unit time.

Q: Highest stock level at the beginning of the cycle.

$TAC(t_w, t_1)$: Total average cost per unit.

$\tilde{c}_1$: Fuzzy fixed cost.

$\tilde{c}_2$: Fuzzy shortage cost per unit per unit time.

$\tilde{c}_3$: Fuzzy deteriorating cost per unit per unit time.

$\tilde{c}_4$: Fuzzy holding cost per unit per unit time.

$\widetilde{TAC}(t_w, t_1)$: Fuzzy total cost per unit.

7.1.2 Assumptions

a) The inventory system involves only one item.

b) The replenishment occurs instantaneously at infinite rate.

c) The lead time is negligible.

d) Demand rate is time depended, we assume it $(a+bt)$ in $0 \le t \le t_w$ and $(c-dt)$ in $t_w \le t \le t_1$.

e) $2\theta t$ is deterioration rate per unit time per cycle, "θ" is constant.

g) At the time $t = t_w$, $(a+bt_w) = (c-dt_w)$.

7.1.3 Crisp inventory model

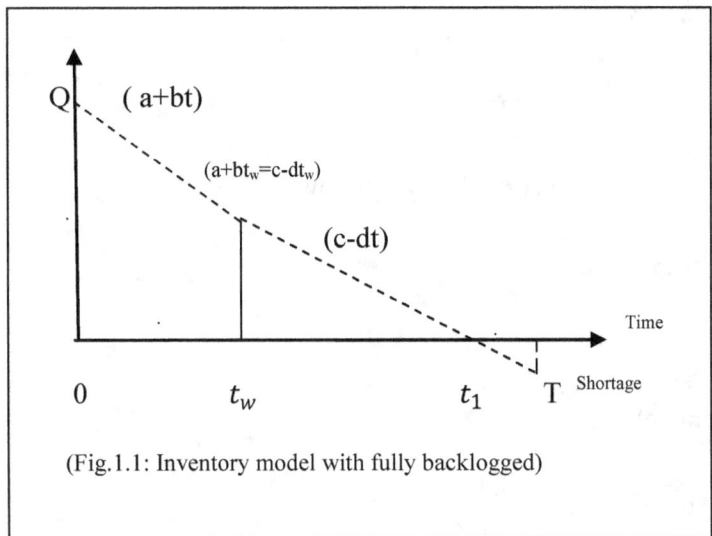

(Fig.1.1: Inventory model with fully backlogged)

During $0 \leq t \leq t_w$ the inventory level decrease due to customer demand (rate of demand $= a+bt$) and deteriorating items. During $t_w \leq t \leq t_1$ the inventory level decreases due to customer demand (rate of demand $(c-dt)$), deteriorating items and reaches to zero at $t = t_1$. In the time interval $t_1 \leq t \leq T$ shortages with fully backlogged allowed.

The differential equations describing the above model as follows

$$\frac{dI(t)}{dt} + 2\theta t I(t) = -(a + bt) \quad , \qquad (0 \leq t \leq t_w). \tag{7.1.3.1}$$

With boundary condition I(0)=Q.

$$\frac{dI(t)}{dt} + 2\theta t I(t) = -(c - dt) \quad , \qquad (t_w \leq t \leq t_1). \tag{7.1.3.2}$$

With boundary condition $I(t_1)$=0

$$\frac{dI(t)}{dt} = -(c - dt), \qquad (t_1 \leq t \leq T). \tag{7.1.3.3}$$

With boundary condition $I(t_1)$=0.

Solving (7.1.3.1) we get,

$$I(t) = Q(1 - \theta t^2) - (1 - \theta t^2)(at + \frac{bt^2}{2}). \tag{7.1.3.4}$$

Solving (7.1.3.2) we get,

$$I(t) = c(1 - \theta t^2)(t_1 - t) - \frac{d}{2}(t_1{}^2 - t^2)(1 - \theta t^2). \tag{7.1.3.5}$$

Solving (7.1.3.3) we get,

$$I(t) = c(t_1 - t) - \frac{d}{2}(t_1{}^2 - t^2).$$

(7.1.3.6)

Here θ is too small, i.e., neglecting the higher power of "θ".

The fixed cost per cycle is,

FC = c_1.

Shortage cost per cycle is,

$$Sc = -c_2 \int_{t_1}^{T} I(t)\,dt$$

$$= -c_2[c(t_1 T - \frac{T^2}{2}) - \frac{Ct_1{}^2}{2} - \frac{d}{2}(t_1{}^2 T - \frac{T^3}{3}) + \frac{dt_1{}^3}{3}].$$

Deterioration cost per cycle is,

$$DC = c_3 \int_0^{t_w}(a+bt)I(t)dt + c_3 \int_{t_w}^{t_1}(c - dt)I(t)dt$$

$$= c_3 \left[Q a t_w - \frac{Qa\theta t_w{}^3}{3} - \frac{a^2 t_w{}^2}{2} - \frac{ab t_w{}^3}{6} + \frac{a^2 \theta t_w{}^4}{4} + \frac{ab\theta t_w{}^5}{10} + \frac{bQ t_w{}^2}{2} - \frac{Qb\theta t_w{}^4}{4} - \frac{ab t_w{}^3}{3}\frac{b^2 t_w{}^4}{8} + \right.$$

$$\frac{ab\theta t_w{}^5}{5} + \frac{b^2\theta t_w{}^6}{12} + c^2\left\{t_1(t_1 - t_w) - \frac{t_1{}^2 - t_w{}^2}{2} - \frac{\theta t_1(t_1{}^3 - t_w{}^3)}{3} + \frac{\theta(t_1{}^4 - t_w{}^4)}{4}\right\} -$$

$$cd\left\{\frac{t_1(t_1{}^2 - t_w{}^2)}{2}\frac{(t_1{}^3 - t_w{}^3)}{3}\frac{\theta t_1(t_1{}^4 - t_w{}^4)}{4} + \frac{\theta(t_1{}^5 - t_w{}^5)}{5}\right\} - \frac{cd}{2}\left\{t_1{}^2(t_1 - t_w) - \frac{(t_1{}^3 - t_w{}^3)}{3} - \right.$$

$$\left. \frac{\theta t_1{}^2(t_1{}^3 - t_w{}^3)}{3} - \frac{\theta(t_1{}^5 - t_w{}^5)}{5}\right\} + \frac{d^2}{2}\left\{\frac{t_1{}^2(t_1{}^2 - t_w{}^2)}{2} - \frac{(t_1{}^4 - t_w{}^4)}{4} - \frac{\theta t_1{}^2(t_1{}^4 - t_w{}^4)}{4} + \frac{\theta(t_1{}^6 - t_w{}^6)}{6}\right\}\right].$$

Holding cost per cycle is,

$$HC = c_4 \int_0^{t_w} I(t)\,dt + c_4 \int_{t_w}^{t_1} I(t)\,dt$$

$$= c_4\left[Q\left(t_w - \frac{Q t_w{}^3}{3}\right) - \left(\frac{a t_w{}^2}{2} + \frac{b t_w{}^3}{6}\right) + \theta\left(\frac{a t_w{}^4}{4} + \frac{b t_w{}^5}{10}\right) + c\left\{t_1(t_1 - t_w) - \frac{t_1{}^2 - t_w{}^2}{2}\right\} - \right.$$

$$\left. c\theta\left\{\frac{t_1{}^3(t_1{}^3 - t_w{}^3)}{3} - \frac{(t_1{}^4 - t_w{}^4)}{4}\right\} - \frac{d}{2}\left\{t_1{}^2(t_1 - t_w) - \frac{t_1{}^3 - t_w{}^3}{3}\right\} + \frac{d\theta}{2}\left\{\frac{t_1{}^2(t_1{}^3 - t_w{}^3)}{3} - \frac{t_1{}^5 - t_w{}^5}{5}\right\}\right].$$

So total average cost per cycle is,

$$\text{TAC}(t_1, t_w) = \frac{1}{T}[FC + SC + DC + HC]$$

$$= \frac{1}{T}\Bigg[c_1 - c_2\Big\{c\Big(t_1 T - \frac{T^2}{2}\Big) - \frac{Ct_1^2}{2} - \frac{d}{2}\Big(t_1^2 T - \frac{T^3}{3}\Big) + \frac{dt_1^3}{3}\Big\} + c_3\Big[Q\alpha t_w - \frac{Q a\theta t_w^3}{3} - \frac{a^2 t_w^2}{2} -$$

$$\frac{ab t_w^3}{6} + \frac{a^2\theta t_w^4}{4} + \frac{ab\theta t_w^5}{10} + \frac{bQ t_w^2}{2} - \frac{Q b\theta t_w^4}{4} - \frac{ab t_w^3}{3} - \frac{b^2 t_w^4}{8} + \frac{ab\theta t_w^5}{5} + \frac{b^2\theta t_w^6}{12} + c^2\Big\{t_1(t_1 -$$

$$t_w) - \frac{t_1^2 - t_w^2}{2} - \frac{\theta t_1(t_1^3 - t_w^3)}{3} + \frac{\theta(t_1^4 - t_w^4)}{4}\Big\} - cd\Big\{\frac{t_1(t_1^2 - t_w^2)}{2} - \frac{(t_1^3 - t_w^3)}{3} - \frac{\theta t_1(t_1^4 - t_w^4)}{4} +$$

$$\frac{\theta(t_1^5 - t_w^5)}{5}\Big\} - \frac{cd}{2}\Big\{t_1^2(t_1 - t_w) - \frac{(t_1^3 - t_w^3)}{3} - \frac{\theta t_1^2(t_1^3 - t_w^3)}{3} - \frac{\theta(t_1^5 - t_w^5)}{5}\Big\} + \frac{d^2}{2}\Big\{\frac{t_1^2(t_1^2 - t_w^2)}{2} -$$

$$\frac{(t_1^4 - t_w^4)}{4} - \frac{\theta t_1^2(t_1^4 - t_w^4)}{4} + \frac{\theta(t_1^6 - t_w^6)}{6}\Big\}\Big] + c_4\Big[Q\Big(t_w - \frac{Q t_w^3}{3}\Big) - \Big(\frac{a t_w^2}{2} + \frac{b t_w^3}{6}\Big) + \theta\Big(\frac{a t_w^4}{4} +$$

$$\frac{b t_w^5}{10}\Big) + c\Big\{t_1(t_1 - t_w) - \frac{t_1^2 - t_w^2}{2}\Big\} - c\theta\Big\{\frac{t_1^3(t_1^3 - t_w^3)}{3} - \frac{(t_1^4 - t_w^4)}{4}\Big\} - \frac{d}{2}\Big\{t_1^2(t_1 - t_w) -$$

$$\frac{t_1^3 - t_w^3}{3}\Big\} + \frac{d\theta}{2}\Big\{\frac{t_1^2(t_1^3 - t_w^3)}{3} - \frac{t_1^5 - t_w^5}{5}\Big\}\Big]\Bigg].$$

According to necessary condition for minimization problem, we must have

$$\frac{\partial TAC(t_w, t_1)}{\partial t_w} = 0, \quad \frac{\partial TAC(t_w, t_1)}{\partial t_1} = 0.$$

From sufficient condition it should satisfies

$$\frac{\partial^2 TAC(t_w, t_1)}{\partial t_w^2} > 0, \quad \frac{\partial^2 TAC(t_w, t_1)}{\partial t_1^2} > 0, \text{ and } \Big[\frac{\partial^2 TAC(t_w, t_1)}{\partial t_w^2}\Big]\Big[\frac{\partial^2 TAC(t_w, t_1)}{\partial t_w^2}\Big] - \Big[\frac{\partial^2 TAC(t_w, t_1)}{\partial t_w.\partial t_1}\Big]^2 > 0.$$

7.1.4 Fuzzy inventory model

Due to uncertainly lets we assume the cost coefficients as a triangular fuzzy number (TFN), as

$$\widetilde{c_1} = (c_1^1, c_1^2, c_1^3), \widetilde{c_2} = (c_2^1, c_2^2, c_2^3), \widetilde{c_3} = (c_3^1, c_3^2, c_3^3), \text{ and } \widetilde{c_4} = (c_4^1, c_4^2, c_4^3).$$

Then the total average cost is given by,

$$\widetilde{TAC}(t_w, t_1) = \frac{1}{T}\Bigg[\widetilde{c_1} - \widetilde{c_2}\Big\{c\Big(t_1 T - \frac{T^2}{2}\Big) - \frac{Ct_1^2}{2} - \frac{d}{2}\Big(t_1^2 T - \frac{T^3}{3}\Big) + \frac{dt_1^3}{3}\Big\} + \widetilde{c_3}\Big[Q\alpha t_w - \frac{Q a\theta t_w^3}{3} -$$

$$\frac{a^2 t_w^2}{2} - \frac{ab t_w^3}{6} + \frac{a^2\theta t_w^4}{4} + \frac{ab\theta t_w^5}{10} + \frac{bQ t_w^2}{2} - \frac{Q b\theta t_w^4}{4} - \frac{ab t_w^3}{3} - \frac{b^2 t_w^4}{8} + \frac{ab\theta t_w^5}{5} + \frac{b^2\theta t_w^6}{12} +$$

$$c^2\Big\{t_1(t_1 - t_w) - \frac{t_1^2 - t_w^2}{2} - \frac{\theta t_1(t_1^3 - t_w^3)}{3} + \frac{\theta(t_1^4 - t_w^4)}{4}\Big\} - cd\Big\{\frac{t_1(t_1^2 - t_w^2)}{2} - \frac{(t_1^3 - t_w^3)}{3} -$$

$$\frac{\theta t_1(t_1^4 - t_w^4)}{4} + \frac{\theta(t_1^5 - t_w^5)}{5}\Big\} - \frac{cd}{2}\Big\{t_1^2(t_1 - t_w) - \frac{(t_1^3 - t_w^3)}{3} - \frac{\theta t_1^2(t_1^3 - t_w^3)}{3} - \frac{\theta(t_1^5 - t_w^5)}{5}\Big\} +$$

$$\frac{d^2}{2}\Big\{\frac{t_1^2(t_1^2 - t_w^2)}{2} - \frac{(t_1^4 - t_w^4)}{4} - \frac{\theta t_1^2(t_1^4 - t_w^4)}{4} + \frac{\theta(t_1^6 - t_w^6)}{6}\Big\}\Big] + \widetilde{c_4}\Big[Q\Big(t_w - \frac{Q t_w^3}{3}\Big) - \Big(\frac{a t_w^2}{2} +$$

$$\frac{b t_w^3}{6}\Big) + \theta\Big(\frac{a t_w^4}{4} + \frac{b t_w^5}{10}\Big) + c\Big\{t_1(t_1 - t_w) - \frac{t_1^2 - t_w^2}{2}\Big\} - c\theta\Big\{\frac{t_1^3(t_1^3 - t_w^3)}{3} - \frac{(t_1^4 - t_w^4)}{4}\Big\} -$$

$$\frac{d}{2}\Big\{t_1^2(t_1 + t_w) - \frac{t_1^3 - t_w^3}{3}\Big\} + \frac{d\theta}{2}\Big\{\frac{t_1^2(t_1^3 - t_w^3)}{3} - \frac{t_1^5 - t_w^5}{5}\Big\}\Big]\Bigg].$$

We defuzzify the fuzzy total cost $\widetilde{TAC}(t_w, t_1)$ by graded mean representation method as follows,

$$\widetilde{TAC}(t_w, t_1) = \frac{1}{6}\left[\widetilde{TAC^1}(t_w, t_1), \widetilde{TAC^2}(t_w, t_1), \widetilde{TAC^3}(t_w, t_1)\right].$$

Where

$$\widetilde{TAC^r}(t_w, t_1) = \frac{1}{T}\left[\widetilde{c_1^r} - \widetilde{c_2^r}\left\{c\left(t_1 T - \frac{T^2}{2}\right) - \frac{Ct_1^2}{2} - \frac{d}{2}\left(t_1^2 T - \frac{T^3}{3}\right) + \frac{dt_1^3}{3}\right\} + \widetilde{c_3^r}\left[Q\alpha t_w - \frac{Qa\theta t_w^3}{3} - \right.\right.$$

$$\frac{a^2 t_w^2}{2} - \frac{abt_w^3}{6} + \frac{a^2\theta t_w^4}{4} + \frac{ab\theta t_w^5}{10} + \frac{bQt_w^2}{2} - \frac{Qb\theta t_w^4}{4} - \frac{abt_w^3}{3} - \frac{b^2 t_w^4}{8} + \frac{ab\theta t_w^5}{5} + \frac{b^2\theta t_w^6}{12} + c^2\left\{t_1(t_1 - \right.$$

$$t_w) - \frac{t_1^2 - t_w^2}{2} - \frac{\theta t_1(t_1^3 - t_w^3)}{3} + \frac{\theta(t_1^4 - t_w^4)}{4}\right\} - cd\left\{\frac{t_1(t_1^2 - t_w^2)}{2} - \frac{(t_1^3 - t_w^3)}{3} - \frac{\theta t_1(t_1^4 - t_w^4)}{4} + \frac{\theta(t_1^5 - t_w^5)}{5}\right\} - $$

$$\frac{cd}{2}\left\{t_1^2(t_1 - t_w) - \frac{(t_1^3 - t_w^3)}{3} - \frac{\theta t_1^2(t_1^3 - t_w^3)}{3} - \frac{\theta(t_1^5 - t_w^5)}{5}\right\} + \frac{d^2}{2}\left\{\frac{t_1^2(t_1^2 - t_w^2)}{2} - \frac{(t_1^4 - t_w^4)}{4} - \frac{\theta t_1^2(t_1^4 - t_w^4)}{4} + \right.$$

$$\left.\frac{\theta(t_1^6 - t_w^6)}{6}\right\}\right] + \widetilde{c_4^r}\left[Q\left(t_w - \frac{Qt_w^3}{3}\right) - \left(\frac{at_w^2}{2} + \frac{bt_w^3}{6}\right) + \theta\left(\frac{at_w^4}{4} + \frac{bt_w^5}{10}\right) - c\theta\left\{\frac{t_1^3(t_1^3 - t_w^3)}{3} - \frac{(t_1^4 - t_w^4)}{4}\right\} + \right.$$

$$c\left\{t_1(t_1 - t_w) - \frac{t_1^2 - t_w^2}{2}\right\} - \frac{d}{2}\left\{t_1^2(t_1 - t_w) - \frac{t_1^3 - t_w^3}{3}\right\} + \frac{d\theta}{2}\left\{\frac{t_1^2(t_1^3 - t_w^3)}{3} - \frac{t_1^5 - t_w^5}{5}\right\}\right]\right].$$

Where $r = 1, 2, 3$ and the grade-mean representation method for triangular fuzzy number (TFN) is

$$\widetilde{TAC}(t_w, t_1) = \frac{1}{6}[\widetilde{TAC^1}(t_w, t_1) + 4\widetilde{TAC^2}(t_w, t_1) + \widetilde{TAC^3}(t_w, t_1)]$$

According to necessary condition for minimization or maximization problem, we must have

$$\frac{\partial TAC(\widetilde{t_w}, t_1)}{\partial t_w} = 0, \quad \frac{\partial TAC(\widetilde{t_w}, t_1)}{\partial t_1} = 0.$$

From sufficient condition it should satisfies

$$\frac{\partial^2 TAC(\widetilde{t_w}, t_1)}{\partial t_w^2} > 0, \quad \frac{\partial^2 TAC(\widetilde{t_w}, t_1)}{\partial t_1^2} > 0.$$

And

$$\left[\frac{\partial^2 TAC(\widetilde{t_w}, t_1)}{\partial t_w^2}\right]\left[\frac{\partial^2 TAC(\widetilde{t_w}, t_1)}{\partial t_1^2}\right] - \left[\frac{\partial^2 TAC(\widetilde{t_w}, t_1)}{\partial t_w \partial t_1}\right]^2 > 0.$$

7.2   Numerical Solution

For crisp model: Let us take the in-put values, as follows

Table 7.1 (Crisp input values)

C_1	C_2	C_3	C_4	θ	Q	a	b	c	d	T
500	1	2	1	0.1	100	5	10	7	10	2

And the out-put values are

Table 7.2 (Output values)

t_w	t_1	$TAC(t_w,t_1)$
0.100	0.266	266.162

For fuzzy model: Let $\tilde{c}_1, \tilde{c}_2, \tilde{c}_3$ and $\tilde{c}_4$ are triangular fuzzy numbers, as follows

$\tilde{c}_1 = (450,500,550)$, $\tilde{c}_2 = (0.9,1.0,1.1)$, $\tilde{c}_3 = (1.8,2.0,2.2)$, $\tilde{c}_4 = (0.9,1.0, 1.1)$.

Then solution of the fuzzy inventory model by graded mean representation is,

(i) When $\tilde{c}_1, \tilde{c}_2, \tilde{c}_3, \tilde{c}_4$ are all triangular fuzzy numbers then,

$TAC(t_w,t_1)=266.124$, $t_w= 0.100$, $t_1= 0.265$.

(ii) When $\tilde{c}_1, \tilde{c}_2, \tilde{c}_3$ are all triangular fuzzy numbers then

$TAC(t_w,t_1)=266.112$, $t_w= 0.100$, $t_1= 0.266$.

(iii) When $\tilde{c}_1, \tilde{c}_2$ are triangular fuzzy numbers then,

$TAC(t_w,t_1)=266.162$ $t_w= 0.100$, $t_1= 0.266$.

(iv) When $\tilde{c}_1$ is the triangular fuzzy numbers then,

$TAC(t_w,t_1) =266.162$, $t_w= 0.100$, $t_1= 0.266$.

7.3 Sensitivity Analysis

We now examine to sensitivity analysis of the optimal solution (OS) of the model for changing optimal cost, keeping the other parameters is unchanged. The initial data are given from the above numerical example.

Table 7.3 (Sensitivity analysis)

Parameters	% of change	$TAC(t_w, t_1)$	t_w	t_1
C_1=250	-50	141.162	0.100	0.266
C_1=375	-25	203.662	0.100	0.266
C_1=500	0	266.162	0.100	0.266
C_1=625	25	328.662	0.100	0.266
C_1=750	50	391.162	0.100	0.266
C_2=0.50	-50	258.081	0.100	0.266
C_2=0.75	-25	262.121	0.100	0.266
C_2=1.00	0	266.162	0.100	0.266
C_2=1.25	25	270.202	0.100	0.266
C_2=1.50	50	274.243	0.100	0.266
C_3=1.00	-50	273.535	0.100	0.227
C_3=1.50	-25	269.855	0.100	0.251
C_3=2.00	0	266.162	0.100	0.266
C_3=2.50	25	262.462	0.100	0.275
C_3=3.00	50	258.758	0.100	0.282
C_4=0.50	-50	264.458	0.100	0.269
C_4=0.75	-25	265.323	0.100	0.268
C_w=1.00	0	266.162	0.100	0.266
C_4=1.25	25	267.000	0.100	0.264
C_4=1.50	50	267.839	0.100	0.262

7.3.1 Effect, for increment parameters-

1) $TAC(t_w,t_1)$ increase rapidly, for increases of c_1.
2) $TAC(t_w,t_1)$ increase slowly, for increases of c_2.
3) $TAC(t_w,t_1)$ decrease slowly, for increases of c_3.
4) $TAC(t_w,t_1)$ increase slowly, for increases of c_4.

7.4   Conclusion

In this chapter, we have presented a real life inventory problem in crisp and fuzzy environment and presented solution along with sensitivity analysis. The inventory model is developed with time depended demand and shortages under fully backlogged. Here demand rate is considered as (a+bt) in $0 \leq t \leq t_w$, and it (c−dt) in $t_w \leq t \leq t_1$.

Here we have considered triangular fuzzy number (TFN) and solved the fuzzy model by fuzzy graded mean integration representation (GMIR) technique. In future, the other type of fuzzy number such as piecewise linear hyperbolic, L-R fuzzy number, trapezoidal fuzzy number (TrFN), pentagonal fuzzy number (PFN) etc. can be considered to construct the fuzzy inventory model. This model is developed under negligible lead time, so consideration of lead time the model should be more realistic and interesting. This model has been developed for single item but multi-items may be more challenging.

Chapter 8

A Bell Shaped Fuzzy Inventory Model and Its Applications Using Possibilistic Approach

Dubois and Prade (1987) introduced the mean value of a fuzzy number as a closed interval bounded by the expectations calculated from its upper and lower distribution functions. Carlsson and Fuller (2001) introduced the notations of lower possibilistic and upper possibilistic mean values and defined the interval-valued possibilistic mean and investigate its relationship to the interval-valued probabilistic mean. They also introduced the notation of crisp possibilistic mean value and crisp possibilistic variance of continuous possibility distributions, which are consistent with the extension principle.

In this chapter, we have proposed a fuzzy inventory model with constant demand, without shortages in fuzzy environment. *Here the parameters of the inventory model expressed as Bell shaped fuzzy number. We have also developed the concept of possibility theory and possibilistic moment generating functions and some statistical concept as mean, variance, standard deviation (SD) in this economic order quantity (EOQ) model. Also some necessary theorems have been derived here. Finally, the model illustrated by numerical examples and applications.*

8.1   Bell Shaped Fuzzy MF

The standard Bell shaped membership function (MF) fuzzy number is represented by $\tilde{A} = $ (x: a, b, c) and the corresponding membership function as follows:

$$\mu_{\tilde{A}}(x) = \frac{1}{1+\left|\frac{x-c}{a}\right|^{2b}} \qquad (8.1.1)$$

Where "c" determines the center of the corresponding membership function, "a" is half width and "b" (together with a) controls the slopes at the crossover points. Figure of Bell shaped membership function as follows;

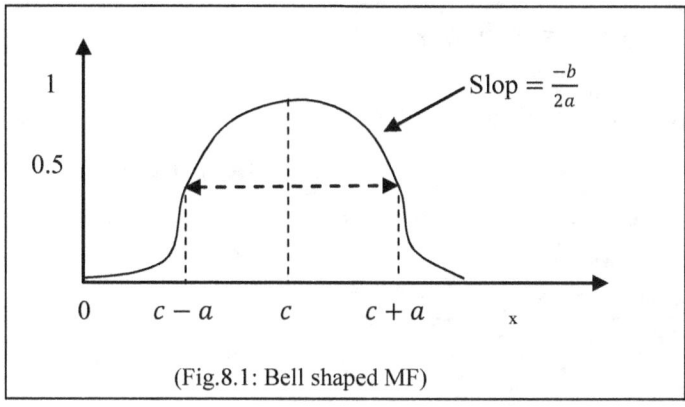

(Fig.8.1: Bell shaped MF)

α-cut of Bell shaped MF defined as follows,

$$\frac{1}{1+\left|\frac{x-c}{a}\right|^{2b}} = \alpha$$

$$x = c \pm a \sqrt[2b]{\frac{1}{\alpha}-1}.$$

So we get $A(\alpha) = [A_1(\alpha), A_2(\alpha)] = \left[c - a\sqrt[2b]{\frac{1}{\alpha}-1},\ c + a\sqrt[2b]{\frac{1}{\alpha}-1}\right].$ \hfill (8.1.2)

Remark 8.1.1: In this paper we developed the model with taking a particular value of b (b = 1).

So (8.1.2) transformed as $A(\alpha) = [A_1(\alpha), A_2(\alpha)] = \left[c - a \times \sqrt{\frac{1}{\alpha}-1},\ c + a \times \sqrt{\frac{1}{\alpha}-1}\right].$

Definition 8.1.2: Let $\hat{A}$ be a fuzzy number, its membership function defined as;

$$\mu_A(x) = \begin{cases} 0 & x \leq a_1 \\ f_A(x) & a_1 \leq x \leq a_2 \\ 1 & a_2 \leq x \leq a_3, \\ g_A(x) & a_3 \leq x \leq a_4 \\ 0 & x \geq a_4 \end{cases}$$ \hfill (8.1.3)

where $f_A(x) : [a_1, a_2] \rightarrow [0, 1]$ is a non decreasing continuous function, $f_A(a_1) = 0$, $f_A(a_2) = 1$, called the left side of the fuzzy number A and $g_A: [a_3, a_4] \rightarrow [0, 1]$ is a non increasing continuous function, $g_A(a_3) = 1$, $g_A(a_4) = 0$, called the right side of the fuzzy number $\hat{A}$.

8.2 Possibilistic Moment Generating Function

8.2.1 Possibilistic mean

According to Appadoo et al., Carlsson and Fuller, we used the following equalities given in (8.2.1) and (8.2.2) deriving some of the results in the EOQ inventory model.

Possibility $[A \leq a_1(\alpha)] = \pi[(-\infty, a_1(\alpha)] = sup_{x \leq a_1(\alpha)}A(x) = \alpha$, $\qquad$ (8.2.1)

Possibility $[A \leq a_2(\alpha)] = \pi[a_2(\alpha), \infty)] = sup_{x \geq a_2(\alpha)}A(x) = \alpha$. $\qquad$ (8.2.2)

For fuzzy number $A(\alpha) = [a_1(\alpha), a_2(\alpha)]$ and $B(\alpha) = [b_1(\alpha), b_2(\alpha)]$, $\alpha \in [0, 1]$, Carlsson and Fuller define crisp lower possibilistic mean value $E_L(A)$, crisp upper possibilistic mean value $E_R(A)$ and crisp possibilistic mean value $E(A)$ as follows

$$E_L(A) = \frac{\int_0^1 Poss\,[A \leq a_1(\alpha)]\min\,[A(\alpha)]d\alpha}{\int_0^1 Poss\,[A \leq a_1(\alpha)]d\alpha} = \frac{\int_0^1 \alpha a_1(\alpha)d\alpha}{\int_0^1 \alpha d\alpha},$$ $\qquad$ (8.2.3)

$$E_R(A) = \frac{\int_0^1 Poss\,[A \geq a_2(\alpha)]\max\,[A(\alpha)]d\alpha}{\int_0^1 Poss\,[A \geq a_2(\alpha)]d\alpha} = \frac{\int_0^1 \alpha a_2(\alpha)d\alpha}{\int_0^1 \alpha d\alpha},$$ $\qquad$ (8.2.4)

and

$$E(A) = \frac{1}{2}\left[\frac{\int_0^1 Poss\,[A \leq a_1(\alpha)]\min\,[A(\alpha)]d\alpha}{\int_0^1 Poss\,[A \leq a_1(\alpha)]d\alpha} + \frac{\int_0^1 Poss\,[A \geq a_2(\alpha)]\max\,[A(\alpha)]d\alpha}{\int_0^1 Poss\,[A \geq a_2(\alpha)]d\alpha}\right]$$

$$= \frac{1}{2}\left[\frac{\int_0^1 \alpha a_1(\alpha)d\alpha}{\int_0^1 \alpha d\alpha} + \frac{\int_0^1 \alpha a_2(\alpha)d\alpha}{\int_0^1 \alpha d\alpha}\right].$$ $\qquad$ (8.2.5)

8.2.2 Possibilistic variance

The lower possibilistic variance($Var_L(A)$), upper possibilistic variance($Var_R(A)$) and possibilistic variance ($Var(A)$) defined as follows;

$$Var_L(A) = \frac{\int_0^1 Poss\,[A \leq a_1(\alpha)](E_L(A) - a_1(\alpha))^2 d\alpha}{\int_0^1 Poss\,[A \leq a_1(\alpha)]d\alpha}$$

$$= 2\int_0^1 \alpha(E_L(A) - a_1(\alpha))^2 d\alpha\,.$$

$$Var_R(A) = \frac{\int_0^1 Poss\,[A \geq a_2(\alpha)](E_L(A) - a_1(\alpha))^2 d\alpha}{\int_0^1 Poss\,[A \geq a_2(\alpha)]d\alpha}$$

$$= 2\int_0^1 \alpha(E_R(A) - a_2(\alpha))^2 d\alpha\,.$$

$$Var(A) = \frac{Var_L(A) + Var_R(A)}{2}$$

$$= \int_0^1 \alpha\{(E_L(A) - a_1(\alpha))^2 + (E_R(A) - a_2(\alpha))^2\}d\alpha.$$

8.2.3 Possibilistic standard deviation (SD)

The non-negative square root of the variance is called standard deviation (Sd).

$$\text{i.e., } Sd(A) = +\sqrt{Var(A)}. \tag{8.2.6}$$

8.3   Deterministic EOQ Model

We have considered an economic lot size model with uniform rate of demand infinite production rate and having no shortages. The notations to be used are;

Tac(Q):Total average cost of the EOQ model.

Q: Order quantity.

c_h: Carrying cost per item per unit time.

c_0: Ordering cost per order.

D: Demand rate per unit time.

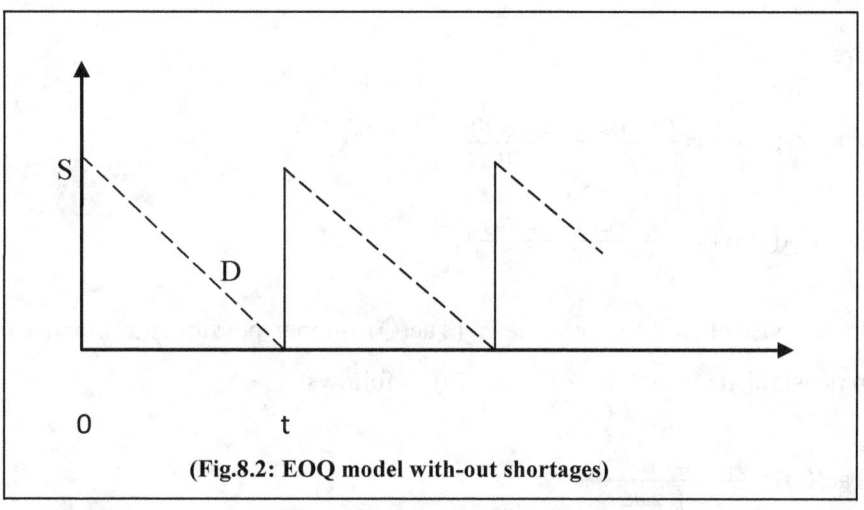

(Fig.8.2: EOQ model with-out shortages)

Variables of this EOQ model only Q and c_0, c_h are constant parameters.

Thus,

Total carrying cost $= \frac{c_h Q t}{2}$,

So total cost $= c_0 + \frac{c_h Q t}{2}$

And total average cost Tac(Q) $= \frac{1}{t}\left[c_0 + \frac{c_h Q t}{2}\right]$

$$= \frac{c_0 D}{Q} + \frac{c_h Q}{2}, \qquad \left[t = \frac{Q}{D}\right]. \tag{8.3.1}$$

Here the optimal order quantity $Q^* = \sqrt{\frac{2c_0 D}{c_h}}. \tag{8.3.2}$

8.4   Fuzzy EOQ Model

In the inventory model we take the parameters c_0 and c_h as fuzzy numbers. Let $C_0(\alpha)$ and $C_h(\alpha)$, denotes the α-cuts of c_0 and c_h respectively. Thus for $0 \leq \alpha \leq 1$ we get

$$C_0(\alpha) = [C_{01}(\alpha), C_{02}(\alpha)],$$

$$C_h(\alpha) = [C_{h1}(\alpha), C_{h2}(\alpha)],$$

Then from (8.3.1) we have

$$\text{Tac}(Q, \alpha) = \frac{[C_{01}(\alpha), C_{02}(\alpha)]D}{Q} + \frac{[C_{h1}(\alpha), C_{h2}(\alpha)]Q}{2}$$

$$= \left[\frac{C_{01}(\alpha)D}{Q} + \frac{C_{h1}(\alpha)Q}{2}, \frac{C_{02}(\alpha)D}{Q} + \frac{C_{h2}(\alpha)Q}{2} \right]$$

$$= [\text{Tac}_1(Q, \alpha), \text{Tac}_2(Q, \alpha)]$$

Where

$$\text{Tac}_1(Q, \alpha) = \frac{C_{01}(\alpha)D}{Q} + \frac{C_{h1}(\alpha)Q}{2Q}, \tag{8.4.1}$$

And $$\text{Tac}_2(Q, \alpha) = \frac{C_{02}(\alpha)D}{Q} + \frac{C_{h2}(\alpha)Q}{2}. \tag{8.4.2}$$

So lower possibilistic mean value $E_L(\text{Tac}(Q))$, upper possibilistic mean value $E_R(\text{Tac}(Q))$ and crisp possibilistic mean value $E(\text{Tac}(Q))$ as follows;

$$E_L(\text{Tac}(Q)) = \frac{\int_0^1 \alpha (\text{Tac}_1(Q, \alpha)) d\alpha}{\int_0^1 \alpha \, d\alpha}$$

$$= 2 \int_0^1 \alpha \left(\frac{C_{01}(\alpha)D}{Q} + \frac{C_{h1}(\alpha)Q}{2} \right) d\alpha, \tag{8.4.3}$$

$$E_R(\text{Tac}(Q)) = \frac{\int_0^1 \alpha (\text{Tac}_2(Q, \alpha)) d\alpha}{\int_0^1 \alpha \, d\alpha}$$

$$= 2 \int_0^1 \alpha \left(\frac{C_{02}(\alpha)D}{Q} + \frac{C_{h2}(\alpha)Q}{2Q} \right) d\alpha, \tag{8.4.4}$$

$$E(\text{Tac}(Q)) = \frac{1}{2} \frac{\int_0^1 \alpha (\text{Tac}_1(Q, \alpha)) d\alpha + \int_0^1 \alpha (\text{Tac}_2(Q, \alpha)) d\alpha}{\int_0^1 \alpha \, d\alpha}$$

$$= \int_0^1 \alpha \left(\frac{C_{02}(\alpha)D}{Q} + \frac{C_{h2}(\alpha)Q}{2Q} + \frac{C_{02}(\alpha)D}{Q} + \frac{C_{h2}(\alpha)Q}{2Q} \right) d\alpha. \tag{8.4.5}$$

So the Interval Valued Possibilistic Mean (IVPM) of Tac(Q,S),

$$= \left[2 \int_0^1 \alpha \left(\frac{C_{01}(\alpha)D}{Q} + \frac{C_{h1}(\alpha)Q}{2Q} \right) d\alpha \,,\, 2 \int_0^1 \alpha \left(\frac{C_{02}(\alpha)D}{Q} + \frac{C_{h2}(\alpha)Q}{2Q} \right) d\alpha \right]. \tag{8.4.6}$$

Theorem 8.4.1

Let the fuzzy set-up cost $c_o(\alpha)$ and fuzzy carrying cost c_h are assume to Bell shaped fuzzy numbers (Remark 8.1.1) and D and Q are crisp order quantities.

Then

i) $\quad E_L(Tac(Q)) = \frac{(8c_{o1} - 2a_{01}\pi)D + (4c_{h1} - a_{h1}\pi)Q^2}{8Q}$;

ii) $\quad E_R(Tac(Q)) = \frac{(8c_{o1} + 2a_{01}\pi)D + (4c_{h1} + a_{h1}\pi)Q^2}{8Q}$,

iii) $\quad E(Tac(Q)) = \frac{(2c_{o1})D + (c_{h1})Q^2}{2Q}$.

Proof

Here, $C_0(\alpha)$, $C_h(\alpha)$, and $C_s(\alpha)$ denotes the Bell shaped α-cuts of c_0 and c_h respectively. Thus for $0 \le \alpha \le 1$ we get,

$$C_0(\alpha) = [C_{01}(\alpha), C_{02}(\alpha)] = \left[c_{01} - a_{01}\sqrt{\frac{1}{\alpha} - 1}\,,\, c_{01} + a_{01}\sqrt{\frac{1}{\alpha} - 1} \right],$$

$$C_h(\alpha) = [C_{h1}(\alpha), C_{h2}(\alpha)] = \left[c_{h1} - a_{h1}\sqrt{\frac{1}{\alpha} - 1}\,,\, c_{h1} + a_{h1}\sqrt{\frac{1}{\alpha} - 1} \right],$$

i) $E_L(Tac(Q)) = \frac{\int_0^1 \alpha a_1(\alpha) d\alpha}{\int_0^1 \alpha d\alpha}$

$$= 2 \int_0^1 \alpha \left\{ \frac{(c_{01} - a_{01}\sqrt{\frac{1}{\alpha} - 1})D}{Q} + \frac{(c_{h1} - a_{h1}\sqrt{\frac{1}{\alpha} - 1})Q}{2} \right\} d\alpha$$

$$= 2 \left[\frac{c_{01}D}{2Q} - \frac{a_{01}D}{Q} \int_0^1 \alpha \times \sqrt{(\frac{1}{\alpha} - 1)}\, d\alpha + \frac{c_{h1}}{4} - \frac{a_{h1}Q}{2} \int_0^1 \alpha \times \sqrt{(\frac{1}{\alpha} - 1)}\, d\alpha \right]$$

$$= \frac{(8c_{o1} - 2a_{01}\pi)D + (4c_{h1} - a_{h1}\pi)Q^2}{8Q}. \tag{8.4.7}$$

ii) $E_R(Tac(Q)) = \frac{\int_0^1 \alpha a_2(\alpha) d\alpha}{\int_0^1 \alpha d\alpha}$

$$= 2\int_0^1 \alpha \left\{ \frac{(c_{01}+a_{01}\sqrt{\frac{1}{\alpha}-1})D}{Q} + \frac{(c_{h1}+a_{h1}\sqrt{\frac{1}{\alpha}-1})Q}{2} \right\} d\alpha$$

$$= 2\left[\frac{c_{01}D}{2Q} + \frac{a_{01}D}{Q}\int_0^1 \alpha \times \sqrt{(\frac{1}{\alpha}-1)}d\alpha + \frac{c_{h1}}{4} + \frac{a_{h1}Q}{2}\int_0^1 \alpha \times \sqrt{(\frac{1}{\alpha}-1)}d\alpha \right]$$

$$= \frac{(8c_{01}+2a_{01}\pi)D+(4c_{h1}+a_{h1}\pi)Q^2}{8Q}. \tag{8.4.8}$$

$$iii)\; E(Tac(Q)) = \frac{1}{2}\left[\frac{\int_0^1 \alpha a_1(\alpha)d\alpha}{\int_0^1 \alpha d\alpha} + \frac{\int_0^1 \alpha a_2(\alpha)d\alpha}{\int_0^1 \alpha d\alpha} \right]$$

$$= \int_0^1 \alpha \left[\frac{(c_{01}-a_{01}\sqrt{\frac{1}{\alpha}-1})D}{Q} + \frac{(c_{h1}-a_{h1}\sqrt{\frac{1}{\alpha}-1})Q}{2} + \frac{(c_{01}+a_{01}\sqrt{\frac{1}{\alpha}-1})D}{Q} + \frac{(c_{h1}+a_{h1}\sqrt{\frac{1}{\alpha}-1})Q}{2} \right] d\alpha$$

$$= \int_0^1 \alpha(\frac{c_{01}D}{Q} + \frac{c_{h1}Q}{2})d\alpha$$

$$= \frac{(2c_{o1})D+(c_{h1})Q^2}{2Q}. \tag{8.4.9}$$

Theorem 8.4.2

For Bell shaped MF, $E(Tac(Q)) = \frac{E_L(Tac(Q))+E_R(Tac(Q))}{2}$, i.e., Crisp possibilistic mean value is average of lower and upper possibilistic mean value.

Proof

$$\frac{E_L(Tac(Q)) + E_R(Tac(Q))}{2}$$

$$= \frac{1}{2}\left[\frac{(8c_{o1}-2a_1\pi)D+(4c_{h1}-a_{h1}\pi)Q^2}{8Q} + \frac{(8c_{o1}+2a_1\pi)D+(4c_{h1}+a_{h1}\pi)Q^2}{8Q} \right]$$

$$= \frac{1}{2}\left[\frac{(8c_{01}-2a_{01}\pi)D+(4c_{h1}-a_{h1}\pi)Q^2+(8c_{01}+2a_{01}\pi)D+(4c_{h1}+a_{h1}\pi)Q^2}{8Q} \right]$$

$$= \frac{1}{2}\left[\frac{16c_{01}D+(8c_{h1})Q^2}{8Q} \right]$$

$$= \frac{2c_{01}D+c_{h1}Q^2}{2Q} = E(Tac(Q)). \tag{8.4.10}$$

Theorem 8.4.3

For optimal value of Q in possibilistic setup form as follows;

$i)\ Q_L{}^* = E_L(Q) = \sqrt{\dfrac{(8c_{01} - 2a_{01}\pi)D}{4c_{h1} - a_{h1}\pi}},$

$ii)\ Q_R{}^* = E_R(Q) = \sqrt{\dfrac{(8c_{01} + 2a_{01}\pi)D}{4c_{h1} + a_{h1}\pi}},$

$iii)\ Q^* = E(Q) = \sqrt{\dfrac{2c_{01}D}{c_{h1}}}.$

Proof

From necessary and sufficient condition for optimal (minimal) Q in possibilistic setup form, we must have $\dfrac{dE(Tac(Q^*))}{dQ^*} = 0$ and $\dfrac{d^2E(Tac(Q^*))}{dQ^{*2}} > 0.$

$i)\ \dfrac{dE_L(Tac(Q_L))}{dQ_L} = 0$

$\Rightarrow\ -\dfrac{(8c_{01} - 2a_{01}\pi)D}{8Q_L{}^2} + \dfrac{4c_{h1} - a_{h1}\pi}{8} = 0$

$\Rightarrow\ \dfrac{-(8c_{01} + 2a_{01}\pi)D + (4c_{h1} + a_{h1}\pi)Q_L{}^2}{8Q_L{}^2} = 0$

$\Rightarrow\ (4c_{h1} - a_{h1}\pi)Q_L{}^2 = (8c_{01} - 2a_{01}\pi)D$

$i.e.,Q_L{}^* = E_L(Q) = \sqrt{\dfrac{(8c_{01} - 2a_{01}\pi)D}{4c_{h1} - a_{h1}\pi}},$ \hfill (8.4.11)

$ii)\ \dfrac{dE_R(Tac(Q_R))}{dQ_R} = 0$

$\Rightarrow\ -\dfrac{(8c_{01} + 2a_{01}\pi)D}{8Q_R{}^2} + \dfrac{4c_{h1} + a_{h1}\pi}{8} = 0$

$\Rightarrow\ \dfrac{-(8c_{01} + 2a_{01}\pi)D + (4c_{h1} + a_{h1}\pi)Q_R{}^2}{8Q_R{}^2} = 0$

$\Rightarrow\ (4c_{h1} + a_{h1}\pi)Q_R{}^2 = (8c_{01} + 2a_{01}\pi)D$

$i.e.,Q_R{}^* = E_R(Q) = \sqrt{\dfrac{(8c_{01} + 2a_{01}\pi)D}{4c_{h1} + a_{h1}\pi}}$ \hfill (8.4.12)

$iii)\ \dfrac{dE(Tac(Q))}{dQ} = 0$

$\Rightarrow\ -\dfrac{c_{01}D}{2Q^2} + \dfrac{c_{h1}}{4} = 0$

$$\Rightarrow \frac{-(2c_{01})D+(c_{h1})Q^2}{4Q^2} = 0$$

$$\Rightarrow (c_{h1})Q^2 = (2c_{01})D$$

$$\text{i.e.,} Q^* = E(Q) = \sqrt{\frac{2c_{01}D}{c_{h1}}}, \tag{8.4.13}$$

Theorem 8.4.4

The optimal value at possibilistic setup form,

i) $E_L(Tac(Q = Q_L{}^*)) = \frac{\sqrt{(4c_{01}-a_{01}\pi)(4c_{h1}-a_{h1}\pi)D}}{4\sqrt{2}}$,

ii) $E_R(Tac(Q = Q_R{}^*)) = \frac{\sqrt{(4c_{01}+a_{01}\pi)(4c_{h1}+a_{h1}\pi)D}}{4\sqrt{2}}$,

iii) $E(Tac(Q = Q^*)) = \sqrt{\frac{c_{01}c_{h1}D}{2}}$.

Proof

i) $E_L(Tac(Q = Q_L{}^*))$

$$= \frac{(8c_{o1} - 2a_1\pi)D + (4c_{h1} - a_{h1}\pi)Q^2}{8Q}$$

$$= \frac{(8c_{o1}-2a_1\pi)D+(4c_{h1}-a_{h1}\pi)\frac{(8c_{01}-2a_{01}\pi)D}{4c_{h1}-a_{h1}\pi}}{8\sqrt{\frac{(8c_{01}-2a_{01}\pi)D}{4c_{h1}-a_{h1}\pi}}}$$

$$= \frac{2(8c_{o1}-2a_1\pi)D}{8\times\sqrt{\frac{(8c_{01}-2a_{01}\pi)D}{4c_{h1}-a_{h1}\pi}}}$$

$$= \frac{\sqrt{(4c_{01}-a_{01})(4c_{h1}-a_{h1}\pi)D}}{4\sqrt{2}}, \tag{8.4.14}$$

ii) $E_R(Tac(Q = Q_R{}^*))$

$$= \frac{(8c_{o1}+2a_1\pi)D+(4c_{h1}+a_{h1}\pi)Q^2}{8Q}$$

$$= \frac{(8c_{o1}+2a_1\pi)D+(4c_{h1}+a_{h1}\pi)\frac{(8c_{01}+2a_{01}\pi)D}{4c_{h1}+a_{h1}\pi}}{8\sqrt{\frac{(8c_{01}+2a_{01}\pi)D}{4c_{h1}+a_{h1}\pi}}}$$

$$= \frac{2(8c_{o1}+2a_1\pi)D}{8\times\sqrt{\frac{(8c_{01}+2a_{01}\pi)D}{4c_{h1}+a_{h1}\pi}}}$$

$$= \frac{\sqrt{(4c_{01}+a_{01})(4c_{h1}+a_{h1}\pi)D}}{4\sqrt{2}},$$

$$(8.4.15)$$

iii) $E(Tac(Q = Q^*))$

$$= \frac{(2c_{o1})D+(c_{h1})Q^2}{4Q}$$

$$= \frac{(2c_{o1})D+c_{h1}\frac{2c_{01}D}{c_{h1}}}{4\times\sqrt{\frac{2c_{01}D}{c_{h1}}}}$$

$$= \frac{4c_{o1}D}{4\times\sqrt{\frac{2c_{01}D}{c_{h1}}}}$$

$$= \sqrt{\frac{c_{01}c_{h1}D}{2}}.$$

$$(8.5.16)$$

Theorem 8.4.5

With the Bell shaped fuzzy MF,

$$Var_L\big(Tac(Q)\big) = Var_R\big(Tac(Q)\big) = Var\big(Tac(Q)\big)$$

$$= \frac{\{64(a_{01})^2\}-4(a_{01})^2\pi^2\}D^2+\{64a_{01}a_{h1}-4a_{01}a_{h1}\pi^2\}DQ^2+\{16(a_{h1})^2\}-(a_{h1})^2\pi^2\}Q^4}{64Q^2},$$

Proof

$$Var_L(Tac(Q)) = 2\int_0^1 \alpha(E_L(Tac(Q)) - a_1(\alpha))^2 d\alpha$$

$$= 2\int_0^1 \alpha a_1(\alpha)^2 d\alpha - E_L(Tac(Q))^2$$

$$= 2\int_0^1 \alpha(\frac{(c_{01}-a_{01}\sqrt{\frac{1}{\alpha}-1})D}{Q} + \frac{(c_{h1}-a_{h1}\sqrt{\frac{1}{\alpha}-1})Q}{2})^2\, d\alpha - (\frac{(8c_{o1}-2a_1\pi)D+(4c_{h1}-a_{h1}\pi)Q^2}{8Q})^2$$

$$= \frac{\{8(c_{01})^2 - 4a_{01}c_{01}\pi + 8(a_{01})^2\}D^2 + (8c_{01}c_{h1} - 2a_{01}c_{h1}\pi - 2a_{h1}c_{01}\pi + 8a_{01}a_{h1})DQ^2 + \{2(c_{h1})^2 - a_{h1}c_{h1}\pi + (a_{h1})^2\}Q^4}{8Q^2}$$

$$- \frac{\{64(c_{01})^2 - 32a_{01}c_{01}\pi + 4(a_{01})^2\}D^2 + (64c_{01}c_{h1} - 16a_{01}c_{h1}\pi - 16a_{h1}c_{01}\pi + 4a_{01}a_{h1}\pi^2)DQ^2 + \{16(c_{h1})^2 - 8a_{h1}c_{h1}\pi + (a_{h1})^2\pi^2\}Q^4}{64Q^2}$$

$$= \frac{\{64(a_{01})^2\} - 4(a_{01})^2\pi^2\}D^2 + \{64a_{01}a_{h1} - 4a_{01}a_{h1}\pi^2\}DQ^2 + \{16(a_{h1})^2\} - (a_{h1})^2\pi^2\}Q^4}{64Q^2}.$$

$$Var_R(Tac(Q)) = 2\int_0^1 \alpha(E_R(Tac(Q)) - a_2(\alpha))^2 d\alpha$$

$$= 2\int_0^1 \alpha a_2(\alpha)^2 d\alpha - E_R(\text{Tac}(Q))^2$$

$$= 2\int_0^1 \alpha\left(\frac{(c_{01}+a_{01}\sqrt{\frac{1}{\alpha}-1})D}{Q} + \frac{(c_{h1}+a_{h1}\sqrt{\frac{1}{\alpha}-1})Q}{2}\right)^2 d\alpha - \left(\frac{(8c_{o1}+2a_1\pi)D+(4c_{h1}+a_{h1}\pi)Q^2}{8Q}\right)^2$$

$$= \frac{\{8(c_{01})^2 + 4a_{01}c_{01}\pi + 8(a_{01})^2\}D^2 + (8c_{01}c_{h1} + 2a_{01}c_{h1}\pi + 2a_{h1}c_{01}\pi + 8a_{01}a_{h1})DQ^2 + \{2(c_{h1})^2 + a_{h1}c_{h1}\pi + (a_{h1})^2\}Q^4}{8Q^2}$$

$$- \frac{\{64(c_{01})^2 + 32a_{01}c_{01}\pi + 4(a_{01})^2\pi^2\}D^2 + (64c_{01}c_{h1} + 16a_{01}c_{h1}\pi + 16a_{h1}c_{01}\pi + 4a_{01}a_{h1}\pi^2)DQ^2 + \{16(c_{h1})^2 + 8a_{h1}c_{h1}\pi + (a_{h1})^2\pi^2\}Q^4}{64Q^2}$$

$$= \frac{\{64(a_{01})^2)-4(a_{01})^2\pi^2\}D^2+\{64a_{01}a_h-4a_{01}a_{h1}\pi^2\}DQ^2+\{16(a_{h1})^2)-(a_{h1})^2\pi^2\}Q^4}{64Q^2}.$$

So $Var_L(Tac(Q)) = Var_R(Tac(Q))$

Now $\text{Var}(Tac(Q)) = \dfrac{Var_L(Tac(Q))+Var_R(Tac(Q))}{2}$

$$= \frac{2Var_L(Tac(Q))}{2}, \qquad [Var_L(Tac(Q)) = Var_R(Tac(Q))].$$

$$= Var_L(Tac(Q)).$$

i.e., $Var_L(Tac(Q)) = Var_R(Tac(Q)) = Var(Tac(Q))$

$$= \frac{\{64(a_{01})^2)-4(a_{01})^2\pi^2\}D^2+\{64a_{01}a_{h1}-4a_{01}a_{h1}\pi^2\}DQ^2+\{16(a_{h1})^2)-(a_{h1})^2\pi^2\}Q^4}{64Q^2}. \qquad (8.4.17)$$

Theorem 8.4.6

For optimal value of Q in possibilistic variance form as follows;

$$Q_V{}^* = Var(Q) = \sqrt{\frac{2a_{01}D}{a_{h1}}}.$$

Proof

From the necessary and sufficient condition for optimal value (minimal value) Q in possibilistic variance form, we must have $\dfrac{dVar(Tac(Q_V))}{dQ_V} = 0$ and $\dfrac{d^2Var(Tac(Q_V))}{dQ_V{}^2} > 0$.

Now $\dfrac{dVar(Tac(Q_V))}{dQ_V} = 0$

$$\Rightarrow \frac{-2\{64(a_{01})^2)-4(a_{01})^2\pi^2\}D^2}{Q_V{}^3} + 2\{16(a_{h1})^2)-(a_{h1})^2\pi^2\}Q_V = 0$$

$$\Rightarrow \{16(a_{h1})^2)-(a_{h1})^2\pi^2\}Q_V = \frac{\{64(a_{01})^2)-4(a_{01})^2\pi^2\}D^2}{Q_V{}^3}$$

$$\Rightarrow Q_V{}^4 = \frac{\{64(a_{01})^2)-4(a_{01})^2\pi^2\}D^2}{16(a_{h1})^2)-(a_{h1})^2\pi^2}$$

$$\Rightarrow Q_V{}^4 = \frac{4(a_{01})^2 D^2}{(a_{h1})^2}$$

i.e.,$Q_V{}^* = Var(Q) = \sqrt{\dfrac{2a_{01}D}{a_{h1}}}$ \hfill (8.4.18)

Theorem 8.4.7

The standard deviation $Sd(Tac(Q))$ defined as follows;

$$Sd(Tac(Q) = \sqrt{\frac{\{64(a_{01})^2)-4(a_{01})^2\pi^2\}D^2+\{64a_1a_h-4a_1a_h\pi^2\}DQ^2+\{16(a_{h1})^2)-(a_{h1})^2\pi^2\}Q^4}{64Q^2}}.$$

Proof

We know $Sd(Tac(Q)) = +\sqrt{Var(Tac(Q))}$.

But $Var\big(Tac(Q)\big) = \dfrac{\{64(a_{01})^2)-4(a_{01})^2\pi^2\}D^2+\{64a_1a_h-4a_1a_h\pi^2\}DQ^2+\{16(a_{h1})^2)-(a_{h1})^2\pi^2\}Q^4}{64Q^2}$

i.e., $(Tac(Q)) = \sqrt{\dfrac{\{64(a_{01})^2)-4(a_{01})^2\pi^2\}D^2+\{64a_1a_h-4a_1a_h\pi^2\}DQ^2+\{16(a_{h1})^2)-(a_{h1})^2\pi^2\}Q^4}{64Q^2}}.$ \hfill (8.4.19)

8.5 Numerical Solution

Let crisp input value of D = 10, and $\mu_{c_0}(x)$, $\mu_{c_h}(x)$ be Bell fuzzy membership functions (MF) (Remark 8.1.1) of fuzzy numbers c_0 and c_h respectively. Where,

$$\mu_{c_0}(x) = \frac{1}{1+\left|\frac{x-100}{75}\right|^2}, \quad \mu_{c_h}(x) = \frac{1}{1+\left|\frac{x-150}{100}\right|^2},$$

$$C_0(\alpha) = [C_{01}(\alpha),C_{02}(\alpha)] = \left[100 - 75 \times \sqrt{\frac{1}{\alpha}-1}, 100 + 75 \times \sqrt{\frac{1}{\alpha}-1}\right],$$

$$C_h(\alpha) = [C_{h1}(\alpha),C_{h2}(\alpha)] = \left[150 - 100 \times \sqrt{\frac{1}{\alpha}-1}, 150 + 100 \times \sqrt{\frac{1}{\alpha}-1}\right],$$

i) Then optimal Q and Tac(Q) in possibilistic setup from (8.4.7) to (8.4.16) are follows,

Table-8.1 (Lower possibilistic out-put values)

$E_L(Q)$	$E_L(Tac(Q_L^*))$
3.391	329.116

Table-8.2 (Crisp possibilistic out-put values)

$E(Q)$	$E(Tac(Q^*))$
3.651	547.723

Table-8.3(Upper possibilistic out-put values)

$E_R(Q)$	$E_R(Tac(Q_R^*))$
3.729	773.354

Here IVPM (Tac(Q,S)) = [329.116,773.354]

ii) The optimal Q_V^* and $Var(Tac(Q))$ from (8.4.17) to (8.4.18) as follows,

Table-8.4(Out-put values of variance)

Q_V^*	$Var(Tac(Q_V^*))$
3.873	57397.960

iv)   The optimal $Sd(Tac(Q))$ from (8.4.19) as follows,

Tabl-8.5 (Out-put values of standard deviation)

$Sd(Tac(Q_V^*))$
239.579

8.6   Sensitivity Analysis

We now examine to sensitivity analysis of the optimal solution of the inventory model for changes of D, keeping the other parameters unchanged.

Table-8.6 (Sensitivity analysis)

D	% of change	$E_L(Q)$	$E(Q)$	$E_R(Q)$	$E_L(Tac(Q_L^*))$	$E(Tac(Q^*))$	$E_R(Tac(Q_R^*))$	$Var(Tac(Q))$	$Sd(Tac(Q))$
2	-80	1.517	1.633	1.668	171.009	328.604	495.608	20663.38	143.748
4	-60	2.145	2.309	2.358	208.840	383.384	565.044	28125.10	167.705

6	-40	2.627	2.828	2.889	246.671	438.163	634.481	36734.77	191.663
8	-20	3.033	3.266	3.335	284.502	492.943	703.917	46492.39	215.621
10	0	3.391	3.651	3.729	329.116	547.723	773.354	57397.96	239.579
12	+20	3.715	4.000	4.085	360.164	602.502	842.790	69451.48	263.536
14	+40	4.012	4.320	4.412	397.995	657.282	912.227	82652.94	287.494
16	+60	4.290	4.619	4.717	435.826	712.061	981.663	97002.36	311.452
18	+80	4.550	4.899	5.003	473.657	766.841	1051.10	112499.7	335.410

Effect, for increment parameters-:

From the table (Table-8.6) we see that all of decision variables increase for increasing of D.

8.7   Conclusion

In this chapter we have developed a Bell shaped fuzzy approach to the EOQ model with constant demand, without shortages. Here we have discussed only a particular type of Bell shape membership function (MF) (taking b = 1). In future general form of Bell shape MF would be. The methodology proposed in this chapter may also be applicable to other economic order quantity (EOQ) models. Our approach provide here a simple EOQ model, but in future it should be used many complex EOQ model. For future research of uncertainty in EOQ model, by using different type of fuzzy numbers such as random fuzzy number, institutional fuzzy number or adaptive fuzzy demand rate be analytically more challenging and interesting.

Chapter 9

A Bell Shaped Fuzzy Economic Order Quantity (EOQ) Model with Constant Demand, Shortages under Fully Backlogged

Carlsson and Fuller (2001) introduced possibilistic moment generating function of fuzzy numbers. Apadoo et al. (2008) extended some of those results to economic order quantity inventory model (EOQ) using nonlinear type of fuzzy numbers. They combined fuzzy technique assisted by possibility theory to deal with uncertainty and derive some important results. The concept of possibilistic mean, variance, covariance and slandered deviation is used in many different areas especially in production management by many different researcher and authors.

In this chapter, we have proposed an inventory model with shortages under fully backlogging in fuzzy environment. Here we have expressed Bell shaped fuzzy number for the parameters of the model. Here we have also developed the concept of possibility theory and possibilistic moment generating function. Three type of possibilistic mean values as Lower possibilistic $(E_L(A))$, Upper possibilistic $(E_R(A))$ and Crisp possibilistic $(E(A))$ of total average cost function developed here. Also some necessary theorems have been derived here. Finally, the model illustrated by numerical examples and applications.

9.1 Bell Shaped Fuzzy MF

The standard Bell shaped membership function (MF) fuzzy number is represented by $\tilde{A} = (x: a, b, c)$ and the corresponding membership function is defined as;

$$\mu_{\tilde{A}}(x) = \frac{1}{1+\left|\frac{x-c}{a}\right|^{2b}} \tag{9.1.1}$$

Where "c" determines the center of the corresponding membership function, "a" is half width and "b" (together with a) controls the slopes at the crossover points. Figure of Bell shaped membership function as follows;

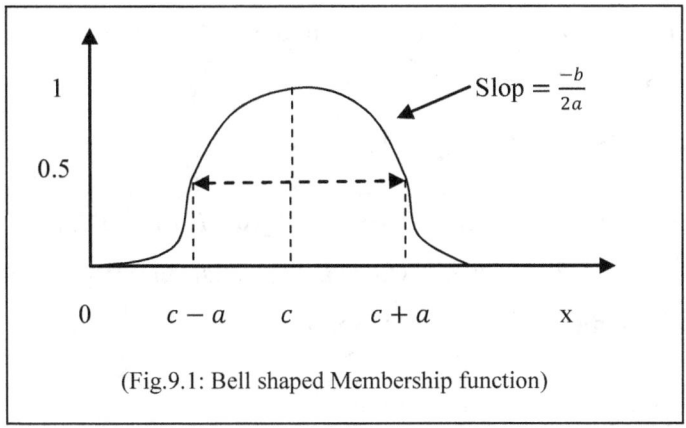

(Fig.9.1: Bell shaped Membership function)

α-cut of Bell shaped MF defined as follows,

$$\frac{1}{1+\left|\frac{x-c}{a}\right|^{2b}} = \alpha$$

Or $x = c \pm a\sqrt[2b]{\frac{1}{\alpha} - 1}$.

So we get $A(\alpha) = [A_1(\alpha), A_2(\alpha)] = \left[c - a\sqrt[2b]{\frac{1}{\alpha} - 1},\ c + a\sqrt[2b]{\frac{1}{\alpha} - 1}\right].$ \hfill (9.1.2)

Remark 9.1.1: In this paper we developed the model with taking a particular value of b (b = 1).

So (9.1.2) transformed as $A(\alpha) = [A_1(\alpha), A_2(\alpha)] = \left[c - a \times \sqrt{\frac{1}{\alpha} - 1},\ c + a \times \sqrt{\frac{1}{\alpha} - 1}\right].$

Definition 9.1.2 Let $\hat{A}$ is a fuzzy number, its membership function defined as;

$$\mu_A(x) = \begin{cases} 0 & x \leq a_1 \\ f_A(x) & a_1 \leq x \leq a_2 \\ 1 & a_2 \leq x \leq a_3, \\ g_A(x) & a_3 \leq x \leq a_4 \\ 0 & x \geq a_4 \end{cases}$$ \hfill (9.1.3)

where $f_A(x) : [a_1, a_2] \rightarrow [0, 1]$ is a non decreasing continuous function, $f_A(a_1) = 0$, $f_A(a_2) = 1$, called the left side of the fuzzy number A and $g_A : [a_3, a_4] \rightarrow [0, 1]$ is a non increasing continuous function, $g_A(a_3) = 1$, $g_A(a_4) = 0$, called the right side of the fuzzy number $\hat{A}$.

9.2   Possibilistic Moment Generating Functions

9.2.1   Possibilistic mean

According to Appadoo et al., Carlsson and Fuller, following equalities given in (9.2.1) and (9.2.2) which are used for deriving some of the results in the EOQ model.

$$\text{Possibility } [A \leq a_1(\alpha)] = \pi[(-\infty, a_1(\alpha)] = sup_{x \leq a_1(\alpha)} A(x) = \alpha, \tag{9.2.1}$$

$$\text{Possibility } [A \leq a_2(\alpha)] = \pi[a_2(\alpha), \infty)] = sup_{x \geq a_2(\alpha)} A(x) = \alpha. \tag{9.2.2}$$

For fuzzy number $A(\alpha) = [a_1(\alpha), a_2(\alpha)]$ and $B(\alpha) = [b_1(\alpha), b_2(\alpha)]$, $\alpha \in [0, 1]$. Carlsson and Fuller define crisp lower possibilistic mean value $E_L(A)$, crisp upper possibilistic mean value $E_R(A)$ and crisp possibilistic mean value $E(A)$ as follows

$$E_L(A) = \frac{\int_0^1 Poss[A \leq a_1(\alpha)] \min[A(\alpha)] d\alpha}{\int_0^1 Poss[A \leq a_1(\alpha)] d\alpha} = \frac{\int_0^1 \alpha a_1(\alpha) d\alpha}{\int_0^1 \alpha d\alpha}, \tag{9.2.3}$$

$$E_R(A) = \frac{\int_0^1 Poss[A \geq a_2(\alpha)] \max[A(\alpha)] d\alpha}{\int_0^1 Poss[A \geq a_2(\alpha)] d\alpha} = \frac{\int_0^1 \alpha a_2(\alpha) d\alpha}{\int_0^1 \alpha d\alpha}, \tag{9.2.4}$$

and
$$E(A) = \frac{1}{2} \left[\frac{\int_0^1 Poss[A \leq a_1(\alpha)] \min[A(\alpha)] d\alpha}{\int_0^1 Poss[A \leq a_1(\alpha)] d\alpha} + \frac{\int_0^1 Poss[A \geq a_2(\alpha)] \max[A(\alpha)] d\alpha}{\int_0^1 Poss[A \geq a_2(\alpha)] d\alpha} \right]$$

$$= \frac{1}{2} \left[\frac{\int_0^1 \alpha a_1(\alpha) d\alpha}{\int_0^1 \alpha d\alpha} + \frac{\int_0^1 \alpha a_2(\alpha) d\alpha}{\int_0^1 \alpha d\alpha} \right]. \tag{9.2.5}$$

9.2.2 Possibilistic variance

The lower possibilistic variance $(Var_L(A))$, upper possibilistic variance $(Var_R(A))$, and possibilistic variance $(Var(A))$ defined as follows;

$$Var_L(A) = \frac{\int_0^1 Poss[A \leq a_1(\alpha)](E_L(A) - a_1(\alpha))^2 d\alpha}{\int_0^1 Poss[A \leq a_1(\alpha)] d\alpha}$$

$$= 2 \int_0^1 \alpha (E_L(A) - a_1(\alpha))^2 d\alpha . \tag{9.2.6}$$

$$Var_R(A) = \frac{\int_0^1 Poss[A \geq a_2(\alpha)](E_L(A) - a_1(\alpha))^2 d\alpha}{\int_0^1 Poss[A \geq a_2(\alpha)] d\alpha}$$

$$= 2 \int_0^1 \alpha (E_R(A) - a_2(\alpha))^2 d\alpha . \tag{9.2.7}$$

$$Var(A) = \frac{Var_L(A) + Var_R(A)}{2}$$

$$= \int_0^1 \alpha \{(E_L(A) - a_1(\alpha))^2 + (E_R(A) - a_2(\alpha))^2\} d\alpha. \tag{9.2.8}$$

9.2.3 Possibilistic standard deviation

The non-negative square root of the variance is called standard deviation (Sd).

i.e., $Sd(A) = +\sqrt{Var(A)}.$ (9.2.9)

9.3 Deterministic EOQ Model

In many real-life situations shortages occur in an EOQ model. So, we have developed an EOQ model with shortage and costs are incurred. The notations to be used are:

Tac(Q,S): Total average cost of the EOQ model.

Q: Order quantity.

$Q - S$:Maximum shortage that occurs under an ordering policy

c_s: Carrying cost per item per unit time.

c_h: Shortages cost per item per unit time.

c_0: Ordering cost per order.

D: Demand rate per unit time.

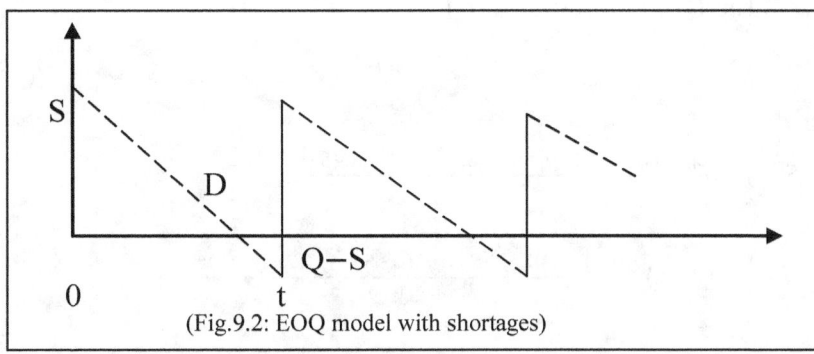

(Fig.9.2: EOQ model with shortages)

Variables of this EOQ model are Q, S and c_0, c_h, c_s are constant parameters.

Thus,

Total carrying cost $= \frac{c_s S^2}{2D}$,

Total shortages cost $= \frac{c_h (q-s)^2}{2D}$,

So total cost $= c_0 + \frac{c_h (q-s)^2}{2D} + \frac{c_s S^2}{2D}$

And total average cost Tac(Q,S) $= \frac{1}{t}\left[c_0 + \frac{c_h (q-s)^2}{2D} + \frac{c_s S^2}{2D}\right]$

$$= \frac{c_0 D}{Q} + \frac{c_h (Q-S)^2}{2Q} + \frac{c_s (S)^2}{2Q}, \qquad \left[t = \frac{Q}{D}\right]. \qquad (9.3.1)$$

Here the optimal order quantity $Q^* = \sqrt{\frac{2c_0 D (c_h + c_s)}{c_h c_s}}, \; S^* = \sqrt{\frac{2c_0 D c_h}{c_s (c_h + c_s)}}.$ (9.3.2)

9.4   Fuzzy EOQ Model

For incurred of costs, we take the cost parameters c_0, c_h and c_s are fuzzy numbers. Let $C_0(\alpha)$, $C_h(\alpha)$, and $C_s(\alpha)$ denotes the α-cuts of c_0, c_h and c_s respectively. Thus for $0 \leq \alpha \leq 1$ we get

$$C_0(\alpha) = [C_{01}(\alpha), C_{02}(\alpha)],$$

$$C_h(\alpha) = [C_{h1}(\alpha), C_{h2}(\alpha)],$$

$$C_s(\alpha) = [C_{s1}(\alpha), C_{s2}(\alpha)].$$

Then from (9.3.1) we have

$$\text{Tac}(Q, S, \alpha) = \frac{[C_{01}(\alpha), C_{02}(\alpha)]D}{Q} + \frac{[C_{h1}(\alpha), C_{h2}(\alpha)](Q-S)^2}{2Q} + \frac{[C_{s1}(\alpha), C_{s2}(\alpha)](S)^2}{2Q}$$

$$= \left[\frac{C_{01}(\alpha)D}{Q} + \frac{C_{h1}(\alpha)(Q-S)^2}{2Q} + \frac{C_{s1}(\alpha)(S)^2}{2Q}, \frac{C_{02}(\alpha)D}{Q} + \frac{C_{h2}(\alpha)(Q-S)^2}{2Q} + \frac{C_{s2}(\alpha)](S)^2}{2Q} \right]$$

$$= [\text{Tac}_1(Q, S, \alpha), \text{Tac}_2(Q, S, \alpha)].$$

Where

$$\text{Tac}_1(Q, S, \alpha) = \frac{C_{01}(\alpha)D}{Q} + \frac{C_{h1}(\alpha)(Q-S)^2}{2Q} + \frac{C_{s1}(\alpha)(S)^2}{2Q}, \tag{9.4.1}$$

And $$\text{Tac}_2(Q, S, \alpha) = \frac{C_{02}(\alpha)D}{Q} + \frac{C_{h2}(\alpha)(Q-S)^2}{2Q} + \frac{C_{s2}(\alpha)(S)^2}{2Q}. \tag{9.4.2}$$

So lower possibilistic mean value $E_L(\text{Tac}(Q,S))$, upper possibilistic mean value $E_R(\text{Tac}(Q,S))$ and crisp possibilistic mean value $E(\text{Tac}(Q,S))$ as follows;

$$E_L(\text{Tac}(Q,S)) = \frac{\int_0^1 \alpha(\text{Tac}_1(Q,S,\alpha))d\alpha}{\int_0^1 \alpha d\alpha}$$

$$= 2\int_0^1 \alpha \left(\frac{C_{01}(\alpha)D}{Q} + \frac{C_{h1}(\alpha)(Q-S)^2}{2Q} + \frac{C_{s1}(\alpha)(S)^2}{2Q} \right)d\alpha, \tag{9.4.3}$$

$$E_R(\text{Tac}(Q,S)) = \frac{\int_0^1 \alpha(\text{Tac}_2(Q,S,\alpha))d\alpha}{\int_0^1 \alpha d\alpha}$$

$$= 2\int_0^1 \alpha \left(\frac{C_{02}(\alpha)D}{Q} + \frac{C_{h2}(\alpha)(Q-S)^2}{2Q} + \frac{C_{s2}(\alpha)(S)^2}{2Q} \right)d\alpha, \tag{9.4.4}$$

$$E(\text{Tac}(Q,S)) = \frac{1}{2} \frac{\int_0^1 \alpha(\text{Tac}_1(Q,S,\alpha))d\alpha + \int_0^1 \alpha(\text{Tac}_2(Q,S,\alpha))d\alpha}{\int_0^1 \alpha d\alpha}$$

$$= \int_0^1 \alpha \left(\frac{C_{02}(\alpha)D}{Q} + \frac{C_{h2}(\alpha)(Q-S)^2}{2Q} + \frac{C_{s2}(\alpha)(S)^2}{2Q} + \frac{C_{02}(\alpha)D}{Q} + \frac{C_{h2}(\alpha)(Q-S)^2}{2Q} + \frac{C_{s2}(\alpha)(S)^2}{2Q} \right) d\alpha. \quad (9.4.5)$$

So the Interval Valued Possibilistic Mean (IVPM) of Tac(Q,S),

$$= \left[2\int_0^1 \alpha \left(\frac{C_{01}(\alpha)D}{Q} + \frac{C_{h1}(\alpha)(Q-S)^2}{2Q} + \frac{C_{s1}(\alpha)(S)^2}{2Q} \right) d\alpha , 2\int_0^1 \alpha \left(\frac{C_{02}(\alpha)D}{Q} + \frac{C_{h2}(\alpha)(Q-S)^2}{2Q} + \frac{C_{s2}(\alpha)(S)^2}{2Q} \right) d\alpha \right] \quad (9.4.6)$$

9.4.1 General case

For $0 \leq \alpha \leq 1$, let the fuzzy set-up cost $c_o(\alpha)$, fuzzy carrying cost c_s and the fuzzy shortages cost c_h be given by

$$C_0(\alpha) = [C_{01}(\alpha), C_{02}(\alpha)] = [c_{01} + \alpha(c_{02} - c_{01}), c_{04} + \alpha(c_{03} - c_{04})],$$

$$C_h(\alpha) = [C_{h1}(\alpha), C_{h2}(\alpha)] = [c_{h1} + \alpha(c_{h2} - c_{h1}), c_{h4} + \alpha(c_{h3} - c_{h4})],$$

$$C_s(\alpha) = [C_{s1}(\alpha), C_{s2}(\alpha)] = [c_{s1} + \alpha(c_{s2} - c_{s1}), c_{s4} + \alpha(c_{s3} - c_{s4})].$$

Then,

Theorem 9.4.1.1

i)$E_L(Tac(Q,S)) = \dfrac{2c_{01}D + 4c_{02}D + (c_{h1} + 2c_{h2})(Q-S)^2 + (c_{s1} + 2c_{s2})S^2}{6Q};$

ii)$E_R(Tac(Q,S)) = \dfrac{4c_{03}D + 2c_{04}D + (c_{h3} + 2c_{h4})(Q-S)^2 + (2c_{s3} + c_{s4})S^2}{6Q},$

iii)$E(Tac(Q,S)) = \dfrac{2c_{01}D + 4c_{02}D + (c_{h1} + 2c_{h2})(Q-S)^2 + (c_{s1} + 2c_{s2})S^2 + 2c_{04}D + 4c_{03}D + (c_{h4} + 2c_{h3})(Q-S)^2 + (c_{s4} + 2c_{s3})S^2}{12Q}.$

Proof

i) $E_L(Tac(Q,S)) = \dfrac{\int_0^1 \alpha a_1(\alpha)d\alpha}{\int_0^1 \alpha d\alpha}$

$$= 2\int_0^1 \alpha \left(\frac{(c_{01} + \alpha(c_{02} - c_{01}))D}{Q} + \frac{(c_{h1} + \alpha(c_{h2} - c_{h1}))(Q-S)^2}{2Q} + \frac{(c_{s1} + \alpha(c_{s2} - c_{s1}))(S)^2}{2Q} \right) d\alpha$$

$$= 2\left[\left(\frac{c_{01}}{2Q} + \frac{c_{02} - c_{01}}{3Q} \right)D + \left(\frac{c_{h1}}{4Q} + \frac{c_{h2} - c_{h1}}{6Q} \right)(Q-S)^2 + \left(\frac{c_{s1}}{4Q} + \frac{c_{s2} - c_{s1}}{6Q} \right)S^2 \right]$$

$$= \frac{2c_{01}D + 4c_{02}D + (c_{h1} + 2c_{h2})(Q-S)^2 + (c_{s1} + 2c_{s2})S^2}{6Q}, \quad (9.4.7)$$

ii) $E_R(Tac(Q,S)) = \dfrac{\int_0^1 \alpha a_2(\alpha)d\alpha}{\int_0^1 \alpha d\alpha}$

$$= 2\int_0^1 \alpha\left(\frac{(c_{04}+\alpha(c_{03}-c_{04}))\,D}{Q} + \frac{(c_{h4}+\alpha(c_{h3}-c_{h4}))(Q-S)^2}{2Q} + \frac{(c_{s4}+\alpha(c_{s3}-c_{s4}))(S)^2}{2Q}\right)d\alpha$$

$$= 2\left[\left(\frac{c_{04}}{2Q} + \frac{c_{03}-c_{04}}{3Q}\right)D + \left(\frac{c_{h4}}{4Q} + \frac{c_{h3}-c_{h4}}{6Q}\right)(Q-S)^2 + \left(\frac{c_{s4}}{4Q} + \frac{c_{s3}-c_{s4}}{6Q}\right)S^2\right]$$

$$= \frac{2c_{04}D + 4c_{03}D + (c_{h4}+2c_{h3})(Q-S)^2 + (c_{s4}+2c_{s3})S^2}{6Q}, \qquad (9.4.8)$$

iii) $E\big(Tac(Q,S)\big) = \frac{1}{2}\left[\dfrac{\int_0^1 \alpha a_1(\alpha)d\alpha}{\int_0^1 \alpha d\alpha} + \dfrac{\int_0^1 \alpha a_2(\alpha)d\alpha}{\int_0^1 \alpha d\alpha}\right]$

$$= \frac{1}{2}\left[\int_0^1 \alpha\left(\frac{(c_{01}+\alpha(c_{02}-c_{01}))\,D}{Q} + \frac{(c_{h1}+\alpha(c_{h2}-c_{h1}))(Q-S)^2}{2Q} + \frac{(c_{s1}+\alpha(c_{s2}-c_{s1}))(S)^2}{2Q}\right)d\alpha + \int_0^1 \alpha\left(\frac{(c_{04}+\alpha(c_{03}-c_{04}))D}{Q} + \right.\right.$$

$$\left.\left.\frac{(c_{h4}+\alpha(c_{h3}-c_{h4}))(Q-S)^2}{2Q} + \frac{(c_{s4}+\alpha(c_{s3}-c_{s4}))(S)^2}{2Q}\right)d\alpha\right]$$

$$= \frac{2c_{01}D+4c_{02}D+(c_{h1}+2c_{h2})(Q-S)^2+(c_{s1}+2c_{s2})S^2+2c_{04}D+4c_{03}D+(c_{h4}+2c_{h3})(Q-S)^2+(c_{s4}+2c_{s3})S^2}{12Q}. \qquad (9.4.9)$$

Theorem 9.4.1.2

For optimal value of Q and S in possibilistic setup form as follows;

i) $\quad Q_L{}^* = E_L(Q) = \sqrt{\dfrac{(2c_{01}D+4c_{02}D)(c_{h1}+2c_{h2}+c_{s1}+2c_{s2})}{(c_{h1}+2c_{h2})(c_{s1}+2c_{s2})}}$,

$\quad S_L{}^* = E_L(S) = \sqrt{\dfrac{(2c_{01}D+4c_{02}D)(c_{h1}+2c_{h2})}{(c_{s1}+2c_{s2})(c_{h1}+2c_{h2}+c_{s1}+2c_{s2})}}$,

ii) $\quad Q_R{}^* = E_R(Q) = \sqrt{\dfrac{(4c_{03}D+2c_{04}D)(2c_{h3}+c_{h4}+2c_{s3}+c_{s4})}{(2c_{h3}+c_{h4})(c_{s3}+2c_{s4})}}$,

$\quad S_R{}^* = E_R(S) = \sqrt{\dfrac{(4c_{03}D+2c_{04}D)(2c_{h3}+c_{h4})}{(2c_{s3}+c_{s4})(2c_{h3}+c_{h4}+2c_{s3}+c_{s4})}}$,

iii) $\quad Q^* = E(Q) = \sqrt{\dfrac{(2c_{01}D+4c_{02}D+4c_{03}D+2c_{04}D)(c_{h1}+2c_{h2}+c_{s1}+2c_{s2}+2c_{h3}+c_{h4}+2c_{s3}+c_{s4})}{(c_{h1}+2c_{h2}+2c_{h3}+c_{h4})(c_{s1}+2c_{s2}+2c_{s3}+c_{s4})}}$,

$\quad S^* = E(S) = \sqrt{\dfrac{(2c_{01}D+4c_{02}D+4c_{03}D+2c_{04}D)(c_{h1}+2c_{h2}+c2_{h3}+c_{h4})}{(c_{s1}+2c_{s2}+2c_{s3}+c_{s4})(c_{h1}+2c_{h2}+c_{s1}+2c_{s2}+2c_{h3}+c_{h4}+2c_{s3}+c_{s4})}}$.

Proof

According to necessary condition for minimization problem, we must have $\frac{\partial TAC\,(Q,S)}{\partial Q} = 0$, $\frac{\partial TAC\,(Q,S)}{\partial S} = 0$. And from sufficient condition it should satisfies $\frac{\partial^2 TAC\,(Q,S)}{\partial Q^2} > 0$, $\frac{\partial^2 TAC\,(Q,S)}{\partial S^2} > 0$ and $\left[\frac{\partial^2 TAC\,(Q,S)}{\partial Q^2}\right]\left[\frac{\partial^2 TAC\,(Q,S)}{\partial S^2}\right] - \left[\frac{\partial^2 TAC\,(Q,S)}{\partial Q\partial S}\right]^2 > 0.$

Now

i) $\dfrac{\partial E_L(\mathrm{Tac}\,(Q_L,S_L))}{\partial Q_L} = 0,$

$$\Rightarrow \frac{-(2c_{01}D+4c_{02}D)+(c_{h1}+2c_{h2})(Q_L-S_L)^2-(c_{s1}+2c_{s2})S_L^2}{Q_L^2} = 0, \qquad (9.4.10)$$

And $\dfrac{\partial E_L(\mathrm{Tac}\,(Q_L,S_L))}{\partial S_L} = 0$

$$\Rightarrow S_L = \frac{(c_{h1}+2c_{h2})Q_L}{(c_{h1}+2c_{h2}+c_{s1}+2c_{s2})}, \qquad (9.4.11)$$

Putting the value of S_L in (9.4.10) and simplify we have,

$$Q_L^* = \sqrt{\frac{(2c_{01}D+4c_{02}D)(c_{h1}+2c_{h2}+c_{s1}+2c_{s2})}{(c_{h1}+2c_{h2})(c_{s1}+2c_{s2})}},$$

Now from (9.4.11) we get,

$$S_L^* = \frac{(c_{h1}+2c_{h2})\sqrt{\dfrac{(2c_{01}D+4c_{02}D)(c_{h1}+2c_{h2}+c_{s1}+2c_{s2})}{(c_{h1}+2c_{h2})(c_{s1}+2c_{s2})}}}{(c_{h1}+2c_{h2}+c_{s1}+2c_{s2})}$$

$$= \sqrt{\frac{(2c_{01}D+4c_{02}D)(c_{h1}+2c_{h2})}{(c_{s1}+2c_{s2})(c_{h1}+2c_{h2}+c_{s1}+2c_{s2})}}.$$

ii) $\dfrac{\partial E_R(\mathrm{Tac}\,(Q_R,S_R))}{\partial Q_R} = 0,$

$$\Rightarrow \frac{-(2c_{04}D+4c_{03}D)+(c_{h4}+2c_{h3})(Q_R-S_R)^2-(c_{s4}+2c_{s3})S_R^2}{Q_R^2} = 0. \qquad (9.4.12)$$

And $\dfrac{\partial E_R(\mathrm{Tac}\,(Q_R,S_R))}{\partial S_R} = 0$

$$\Rightarrow S_R = \frac{(c_{h4}+2c_{h3})Q_R^*}{(c_{h4}+2c_{h3}+c_{s4}+2c_{s3})}, \qquad (9.4.13)$$

Putting the value of S_R in (9.4.12) and simplify we have,

$$Q_R^* = \sqrt{\frac{(2c_{04}D+4c_{03}D)(c_{h1}+2c_{h4}+c_{s1}+2c_{s3})}{(c_{h4}+2c_{h3})(c_{s4}+2c_{s3})}},$$

Now from (9.4.13) we get,

$$S_R^* = \frac{(c_{h4}+2c_{h3})\sqrt{\dfrac{(2c_{04}D+4c_{03}D)(c_{h4}+2c_{h3}+c_{s4}+2c_{s3})}{(c_{h4}+2c_{h3})(c_{s3}+2c_{s4})}}}{(c_{h1}+2c_{h2}+c_{s1}+2c_{s2})}$$

$$= \sqrt{\frac{(2c_{04}D+4c_{03}D)(c_{h4}+2c_{h3})}{(c_{s4}+2c_{s3})(c_{h4}+2c_{h3}+c_{s4}+2c_{s3})}}.$$

iii) $\dfrac{\partial E\,(\mathrm{Tac}\,(Q,S))}{\partial Q} = 0$

$$\Rightarrow \frac{-(2c_{01}D+4c_{02}D+2c_{04}D+4c_{03}D)+(c_{h1}+2c_{h2}+c_{h4}+2c_{h3})(Q-S)^2-(c_{s1}+2c_{s2}+c_{s4}+2c_{s3})S^2}{Q^2}=0. \qquad (9.4.14)$$

And $\dfrac{\partial E\,(\mathrm{Tac}\,(Q,S))}{\partial S}=0$

$$\Rightarrow S = \frac{(c_{h1}+2c_{h4}+c_{h4}+2c_{h3})Q_R}{(c_{h1}+2c_{h2}+c_{s1}+2c_{s2}+c_{h4}+2c_{h3}+c_{s4}+2c_{s3})}, \qquad (9.4.15)$$

Putting the value of S in (9.4.14) and simplify we have,

$$Q^* = \sqrt{\frac{(2c_{01}D+4c_{02}D+4c_{03}D+2c_{04}D)(c_{h1}+2c_{h2}+c_{s1}+2c_{s2}+2c_{h3}+c_{h4}+2c_{s3}+c_{s4})}{(c_{h1}+2c_{h2}+2c_{h3}+c_{h4})(c_{s1}+2c_{s2}+2c_{s3}+c_{s4})}},$$

Now from (9.4.15) we get,

$$S^* = \frac{(c_{h1}+2c_{h4}+c_{h4}+2c_{h3})\sqrt{\dfrac{(2c_{01}D+4c_{02}D+4c_{03}D+2c_{04}D)(c_{h1}+2c_{h2}+c_{s1}+2c_{s2}+2c_{h3}+c_{h4}+2c_{s3}+c_{s4})}{(c_{h1}+2c_{h2}+2c_{h3}+c_{h4})(c_{s1}+2c_{s2}+2c_{s3}+c_{s4})}}}{(c_{h1}+2c_{h2}+c_{s1}+2c_{s2}+c_{h4}+2c_{h3}+c_{s4}+2c_{s3})}$$

$$=\sqrt{\frac{(2c_{01}D+4c_{02}D+4c_{03}D+2c_{04}D)(c_{h1}+2c_{h2}+c2_{h3}+c_{h4})}{(c_{s1}+2c_{s2}+2c_{s3}+c_{s4})(c_{h1}+2c_{h2}+c_{s1}+2c_{s2}+2c_{h3}+c_{h4}+2c_{s3}+c_{s4})}}.$$

Theorem 9.4.1.3

The optimal value at possibilistic setup form,

$$E_L(Tac(Q=Q_L{}^*, S=S_L{}^*))$$

$$= \frac{\sqrt{(c_{h1}+2c_{h2})(2c_{01}D+4c_{02}D)}\{(c_{s1}+2c_{s2})(c_{h1}+2c_{h2}+c_{s1}+2c_{s2})+(2c_{h1}+4c_{h2}+c_{s1}+2c_{s2})^2+(c_{h1}+2c_{h2})(c_{s1}+2c_{s2})\}}{6\times\sqrt{(c_{s1}+2c_{s2})(c_{h1}+2c_{h2}+c_{s1}+2c_{s2})^3}},$$

$$E_R(Tac(Q=Q_R{}^*, S=S_R{}^*))$$

$$= \frac{\sqrt{(2c_{h3}+c_{h4})(4c_{03}D+2c_{04}D)}\{(2c_{s3}+c_{s4})(2c_{h3}+c_{h4}+c_{s3}+2c_{s4})+(4c_{h3}+2c_{h4}+2c_{s3}+c_{s4})^2+(2c_{h3}+c_{h4})(2c_{s3}+c_{s4})\}}{6\times\sqrt{(2c_{s3}+c_{s4})(2c_{h3}+c_{h4}+2c_{s3}+c_{s4})^3}},$$

$$E(Tac(Q=Q^*, S=S^*))$$

$$= \left[\sqrt{(c_{h1} + 2c_{h2} + 2c_{h3} + c_{h4})(2c_{01}D + 4c_{02}D + 4c_{03}D + 2c_{04}D)}\{(c_{s1} + 2c_{s2} + 2c_{s3} + c_{s4})(c_{h1} + 2c_{h2} + c_{s1} + 2c_{s2} + 2c_{h3} + c_{h4} + 2c_{s3} + c_{s4}) + (2c_{h1} + 4c_{h2} + c_{s1} + 2c_{s2} + 4c_{h3} + 2c_{h4} + 2c_{s3} + c_{s4})^2 + (c_{h1} + 2c_{h2} + 2c_{h3} + c_{h4})(c_{s1} + 2c_{s2} + 2c_{s3} + c_{s4})\}\right] \times \frac{1}{12 \times \sqrt{(c_{s1} + 2c_{s2} + 2c_{s3} + c_{s4})(c_{h3} + 2c_{h4} + 2c_{s3} + c_{s4})^3}}.$$

Proof

It is trivial case. Putting the values of $Q_L{}^*, S_L{}^*, Q_R{}^*, S_R{}^*, Q^*$ and S^* in (9.4.7), (9.4.8) and (9.4.9) respectively we can proof the theorem easily.

9.4.2 Special case

Let the fuzzy set-up cost $c_o(\alpha)$, fuzzy carrying cost c_s and the fuzzy shortages cost c_h are assume to Bell shaped (Remark 9.2.1) fuzzy numbers and D,Q and S are crisp order quantities. Let $C_0(\alpha)$, $C_h(\alpha)$, and $C_s(\alpha)$ denotes the α-cuts of c_0, c_h and c_s respectively. Thus for $0 \le \alpha \le 1$ we get,

$$C_0(\alpha) = [C_{01}(\alpha), C_{02}(\alpha)] = \left[c_{01} - a_{01}\sqrt{\frac{1}{\alpha} - 1},\ c_{01} + a_{01}\sqrt{\frac{1}{\alpha} - 1}\right],$$

$$C_h(\alpha) = [C_{h1}(\alpha), C_{h2}(\alpha)] = \left[c_{h1} - a_{h1}\sqrt{\frac{1}{\alpha} - 1},\ c_{h1} + a_{h1}\sqrt{\frac{1}{\alpha} - 1}\right],$$

$$C_s(\alpha) = [C_{s1}(\alpha), C_{s2}(\alpha)] = \left[c_{s1} - a_{s1}\sqrt{\frac{1}{\alpha} - 1},\ c_{s1} + a_{s1}\sqrt{\frac{1}{\alpha} - 1}\right].$$

Then,

Theorem 9.4.2.1

i) $E_L(Tac(Q, S)) = \dfrac{(8c_{01} - 2a_{01}\pi)D + (4c_{h1} - a_{h1}\pi)(Q-S)^2 + (4c_{s1} - a_{s1}\pi)S^2}{8Q}$,

ii) $E_R(Tac(Q, S)) = \dfrac{(8c_{01} + 2a_{01}\pi)D + (4c_{h1} + a_{h1}\pi)(Q-S)^2 + (4c_{s1} + a_{s1}\pi)S^2}{8Q}$,

iii) $E(Tac(Q, S)) = \dfrac{2c_{01}D + c_{h1}(Q-S)^2 + c_{s1}S^2}{2Q}$.

Proof

i) $E_L(Tac(Q, S)) = \dfrac{\int_0^1 \alpha\, a_1(\alpha)d\alpha}{\int_0^1 \alpha\, d\alpha}$

$$= 2\int_0^1 \alpha\left\{\frac{(c_{01} - a_{01}\sqrt{\frac{1}{\alpha} - 1})D}{Q} + \frac{(c_{h1} - a_{h1}\sqrt{\frac{1}{\alpha} - 1})(Q-S)^2}{2Q} + \frac{(c_{s1} - a_{s1}\sqrt{\frac{1}{\alpha} - 1})(S)^2}{2Q}\right\}d\alpha$$

$$= 2\left[\left(\frac{c_{01}}{2Q} - \frac{c_{02}\pi}{8Q}\right)D + \left(\frac{c_{h1}}{4Q} - \frac{c_{h2}\pi}{16Q}\right)(Q-S)^2 + \left(\frac{c_{s1}}{4Q} - \frac{c_{s2}\pi}{16Q}\right)S^2\right]$$

$$= \frac{(8c_{01} - 2a_{01}\pi)D + (4c_{h1} - a_{h1}\pi)(Q-S)^2 + (4c_{s1} - a_{s1}\pi)S^2}{8Q}, \tag{9.4.16}$$

ii) $E_R(Tac(Q,S)) = \frac{\int_0^1 \alpha\, a_2(\alpha)d\alpha}{\int_0^1 \alpha\, d\alpha}$

$$= 2\int_0^1 \alpha\left\{\frac{(c_{01} + a_{01}\sqrt{\frac{1}{\alpha}-1})D}{Q} + \frac{(c_{h1} + a_{h1}\sqrt{\frac{1}{\alpha}-1})(Q-S)^2}{2Q} + \frac{(c_{s1} + a_{s1}\sqrt{\frac{1}{\alpha}-1})(S)^2}{2Q}\right\}d\alpha$$

$$= 2\left[\left(\frac{c_{01}}{2Q} + \frac{c_{02}\pi}{8Q}\right)D + \left(\frac{c_{h1}}{4Q} + \frac{c_{h2}\pi}{16Q}\right)(Q-S)^2 + \left(\frac{c_{s1}}{4Q} + \frac{c_{s2}\pi}{16Q}\right)S^2\right]$$

$$= \frac{(8c_{01} - 2a_{01}\pi)D + (4c_{h1} - a_{h1}\pi)(Q-S)^2 + (4c_{s1} - a_{s1}\pi)S^2}{8Q}, \tag{9.4.17}$$

iii) $E(Tac(Q,S)) = \frac{1}{2}\left[\frac{\int_0^1 \alpha\, a_1(\alpha)d\alpha}{\int_0^1 \alpha\, d\alpha} + \frac{\int_0^1 \alpha\, a_2(\alpha)d\alpha}{\int_0^1 \alpha\, d\alpha}\right]$

$$= \left[\int_0^1 \alpha\left\{\frac{(c_{01} - a_{01}\sqrt{\frac{1}{\alpha}-1})D}{Q} + \frac{(c_{h1} - a_{h1}\sqrt{\frac{1}{\alpha}-1})(Q-S)^2}{2Q} + \frac{(c_{s1} - a_{s1}\sqrt{\frac{1}{\alpha}-1})(S)^2}{2Q}\right\}d\alpha\right.$$

$$\left. + \int_0^1 \alpha\left\{\frac{(c_{01} + a_{01}\sqrt{\frac{1}{\alpha}-1})D}{Q} + \frac{(c_{h1} + a_{h1}\sqrt{\frac{1}{\alpha}-1})(Q-S)^2}{2Q} + \frac{(c_{s1} + a_{s1}\sqrt{\frac{1}{\alpha}-1})(S)^2}{2Q}\right\}d\alpha\right]$$

$$= 2\int_0^1 \alpha\left(\frac{c_{01}D}{Q} + \frac{c_{h1}(Q-S)^2}{2Q} + \frac{c_{h1}(Q-S)^2}{2Q}\right)d\alpha$$

$$= \frac{2c_{01}D + c_{h1}(Q-S)^2 + c_{s1}S^2}{2Q}. \tag{9.4.18}$$

Theorem 9.4.2.2

For Bell shaped MF, $E(Tac(Q,S)) = \frac{E_L(Tac(Q,S)) + E_R(Tac(Q,S))}{2}$ i.e., Crisp possibilistic mean value is average of lower and upper possibilistic mean value.

Proof

$\frac{E_L(Tac(Q,S)) + E_R(Tac(Q,S))}{2}$

$$= \frac{1}{2}\left[\frac{(8c_{01} - 2a_{01}\pi)D + (4c_{h1} - a_{h1}\pi)(Q-S)^2 + (4c_{s1} - a_{s1}\pi)S^2}{8Q}\right.$$

$$\left. + \frac{(8c_{01} + 2a_{01}\pi)D + (4c_{h1} + a_{h1}\pi)(Q-S)^2 + (4c_{s1} + \pi)S^2}{8Q}\right]$$

$$= \frac{1}{2} \left[\frac{16c_{01}D + (8c_{h1})(Q-S)^2 + (8c_{s1})(S)^2}{8Q} \right]$$

$$= \frac{2c_{01}D + c_{h1}(Q-S)^2 + c_{s1}S^2}{2Q} = E(Tac(Q,S)). \tag{9.4.19}$$

Theorem 9.4.2.3

For optimal value of Q and S in possibilistic setup form as follows;

i) $\quad Q_L{}^* = E_L(Q) = \sqrt{\dfrac{(8c_{01}D - 2a_{01}\pi D)(4c_{h1} - a_{h1}\pi + 4c_{s1} - a_{s1}\pi)}{(4c_{h1} - a_{h1}\pi)(4c_{s1} - a_{s1}\pi)}},$

$\quad S_L{}^* = E_L(S) = \sqrt{\dfrac{(8c_{01}D - 2a_{01}\pi D)(4c_{h1} - a_{h1}\pi)}{(4c_{s1} - a_{s1}\pi)(4c_{h1} - a_{h1}\pi + 4c_{s1} - a_{s1}\pi)}},$

ii) $\quad Q_R{}^* = E_R(Q) = \sqrt{\dfrac{(8c_{01}D + 2a_{01}\pi D)(4c_{h1} + a_{h1}\pi + 4c_{s1} + a_{s1}\pi)}{(4c_{h1} + a_{h1}\pi)(4c_{s1} + a_{s1}\pi)}},$

$\quad S_R{}^* = E_R(S) = \sqrt{\dfrac{(8c_{01}D + 2a_{01}\pi D)(4c_{h1} + a_{h1}\pi)}{(4c_{s1} + a_{s1}\pi)(4c_{h1} + a_{h1}\pi + 4c_{s1} + a_{s1}\pi)}},$

iii) $\quad Q^* = E(Q) = \sqrt{\dfrac{(2c_{01}D)(c_{h1} + c_{s1})}{(c_{h1})(c_{s1})}},$

$\quad S^* = E(S) = \sqrt{\dfrac{(2c_{01}D)(c_{h1})}{(c_{s1})(c_{h1} + c_{s1})}}.$

Proof

According to necessary condition for optimal value (minimum value), we must have $\frac{\partial TAC\,(Q,S)}{\partial Q} = 0$, $\frac{\partial TAC\,(Q,S)}{\partial S} = 0$. And from sufficient condition it should satisfies $\frac{\partial^2 TAC\,(Q,S)}{\partial Q^2} > 0$,

$\frac{\partial^2 TAC\,(Q,S)}{\partial S^2} > 0$ and $\left[\frac{\partial^2 TAC\,(Q,S)}{\partial Q^2} \right] \left[\frac{\partial^2 TAC\,(Q,S)}{\partial S^2} \right] - \left[\frac{\partial^2 TAC\,(Q,S)}{\partial Q \partial S} \right]^2 > 0.$

Now

i) $\dfrac{dE_L(Tac(Q_L,S_L))}{dQ_L} = 0$

$\Rightarrow \dfrac{-(8c_{01} - 2a_{01}\pi)D + (4c_{h1} - a_{h1}\pi)(Q_L - S_L)^2 - (4c_{s1} - a_{s1}\pi)S_L{}^2}{Q_L{}^2} = 0 \tag{9.4.20}$

And $\dfrac{\partial E_L(Tac(Q_L,S_L))}{\partial S_L} = 0$

$\Rightarrow S_L = \dfrac{(4c_{h1} - a_{h1}\pi)Q_L}{(4c_{h1} - a_{h1}\pi + 4c_{s1} - a_{s1}\pi)}, \tag{9.4.21}$

Putting the value of S_L in (9.4.20) and simplify we have,

$$Q_L^* = \sqrt{\frac{(8c_{01}D - 2a_{01}\pi D)(4c_{h1} - a_{h1}\pi + 4c_{s1} - a_{s1}\pi)}{(4c_{h1} - a_{h1}\pi)(4c_{s1} - a_{s1}\pi)}},$$

Now from (9.4.21) we get,

$$S_L^* = \frac{(4c_{h1} - a_{h1}\pi)\sqrt{\frac{(8c_{01}D - 2a_{01}\pi D)(4c_{h1} - a_{h1}\pi + 4c_{s1} - a_{s1}\pi)}{(4c_{h1} - a_{h1}\pi)(4c_{s1} - a_{s1}\pi)}}}{(4c_{h1} - a_{h1}\pi + 4c_{s1} - a_{s1}\pi)}$$

$$= \sqrt{\frac{(8c_{01}D - 2a_{01}\pi D)(4c_{h1} - a_{h1}\pi)}{(4c_{s1} - a_{s1}\pi)(4c_{h1} - a_{h1}\pi + 4c_{s1} - a_{s1}\pi)}}.$$

ii) $\dfrac{dE_R(Tac(Q_R, S_R))}{dQ_R} = 0$

$$\Rightarrow \frac{-(8c_{01} + 2a_{01}\pi)D + (4c_{h1} + a_{h1}\pi)(Q_R - S_R)^2 - (4c_{s1} + a_{s1}\pi)S_R^2}{Q_R^2} = 0 \tag{9.4.22}$$

And $\dfrac{\partial E_R(Tac(Q_R, S_R))}{\partial S_R} = 0$

$$\Rightarrow S_R = \frac{(4c_{h1} + a_{h1}\pi)Q_R}{(4c_{h1} + a_{h1}\pi + 4c_{s1} + a_{s1}\pi)}, \tag{9.4.23}$$

Putting the value of S_R in (9.4.22) and simplify we have,

$$Q_R^* = \sqrt{\frac{(8c_{01}D + 2a_{01}\pi D)(4c_{h1} + a_{h1}\pi + 4c_{s1} + a_{s1}\pi)}{(4c_{h1} + a_{h1}\pi)(4c_{s1} + a_{s1}\pi)}},$$

Now from (9.4.23) we get,

$$S_R^* = \frac{(4c_{h1} + a_{h1}\pi)\sqrt{\frac{(8c_{01}D + 2a_{01}\pi D)(4c_{h1} + a_{h1}\pi + 4c_{s1} + a_{s1}\pi)}{(4c_{h1} + a_{h1}\pi)(4c_{s1} + a_{s1}\pi)}}}{(4c_{h1} - a_{h1}\pi + 4c_{s1} - a_{s1}\pi)}$$

$$= \sqrt{\frac{(8c_{01}D + 2a_{01}\pi D)(4c_{h1} + a_{h1}\pi)}{(4c_{s1} + a_{s1}\pi)(4c_{h1} + a_{h1}\pi + 4c_{s1} + a_{s1}\pi)}}.$$

iii) $\dfrac{dE(Tac(Q, S))}{dQ} = 0$

$$\Rightarrow \frac{-2c_{01}D + (c_{h1})(Q - S)^2 - (c_{s1})S^2}{2Q^{*2}} = 0 \tag{9.4.24}$$

And $\dfrac{\partial E(Tac(Q, S))}{\partial S} = 0$

$$\Rightarrow S^* = \frac{(c_{h1})Q}{(c_{h1} + 4c_{s1})}, \tag{9.4.25}$$

Putting the value of S in (9.4.24) and simplify we have,

$$Q^* = \sqrt{\frac{(2c_{01}D)(c_{h1}+c_{s1})}{(c_{h1})(c_{s1})}},$$

Now from (9.4.25) we get,

$$S^* = \frac{(c_{h1})\sqrt{\frac{(2c_{01}D)(c_{h1}+c_{s1})}{(c_{h1})(c_{s1})}}}{(c_{h1}+4c_{s1})}$$

$$= \sqrt{\frac{(2c_{01}D)(c_{h1})}{(c_{s1})(c_{h1}+c_{s1})}}.$$

Theorem 9.4.2.4

The optimal value at possibilistic setup form,

$$E_L(Tac(Q = Q_L^*))$$

$$= \frac{\sqrt{(4c_{h1}D-a_{h1}\pi D)(8c_{01}-2a_{01}\pi)}\{(4c_{s1}-a_{s1}\pi)(4c_{h1}-a_{h1}\pi+4c_{s1}-a_{s1}\pi)+((8c_{h1}-2a_{h1}\pi+4c_{s1}-a_{s1}\pi)^2+(4c_{h1}-a_{h1}\pi)(4c_{s1}-a_{s1})\}}{8\times\sqrt{(4c_{s1}-a_{s1})(4c_{h1}-a_{h1}\pi+4c_{s1}-a_{s1})^3}},$$

$$E_R(Tac(Q = Q_R^*))$$

$$= \frac{\sqrt{(4c_{h1}D+a_{h1}\pi D)(8c_{01}+2a_{01}\pi)}\{(4c_{s1}+a_{s1}\pi)(4c_{h1}+a_{h1}\pi+4c_{s1}+a_{s1}\pi)+((8c_{h1}+2a_{h1}\pi+4c_{s1}+a_{s1}\pi)^2+(4c_{h1}+a_{h1}\pi)(4c_{s1}+a_{s1})\}}{8\times\sqrt{(4c_{s1}+a_{s1})(4c_{h1}+a_{h1}\pi+4c_{s1}+a_{s1})^3}},$$

$$E(Tac(Q = Q^*))$$

$$= \frac{\sqrt{(c_{h1}D)(c_{01})}\{(c_{s1})(c_{h1}+c_{s1})+((2c_{h1}+c_{s1})^2+(c_{h1})(c_{s1})\}}{\sqrt{2}\times\sqrt{(c_{s1})(c_{h1}+c_{s1})^3}}.$$

Proof

It is trivial case. Putting the values of $Q_L^*, S_L^*, Q_R^*, S_R^*, Q^*$ and S^* in (9.4.16), (9.4.17) and (9.4.18) respectively we can proof the theorem easily.

Theorem 9.4.2.5

With the Bell shaped fuzzy MF,

$$Var_L(Tac(Q,S)) = Var_R(Tac(Q,S)) = Var(Tac(Q,S))$$

$$= \frac{1}{64Q^2}\left[64(a_{01})^2) - 4(a_{01})^2\pi^2\}D^2 + \{16(a_{h1})^2) - (a_{h1})^2\pi^2\}(Q-S)^4 + \{16(a_{s1})^2) - (a_{s1})^2\pi^2\}S^4 + \right.$$

$$\left. \{64a_{01}a_{h1} - 4a_{01}a_{h1}\pi^2\}D(Q-S)^2 + \{32a_{h1}a_{s1} - 2a_{h1}a_{s1}\pi^2\}(Q-S)^2S^2 + \{64a_{01}a_{s1} - 4a_{01}a_{s1}\pi^2\}DS^2\right]$$

Proof

$$Var_L(Tac(Q,S)) = 2\int_0^1 \alpha(E_L(Tac(Q,S)) - a_1(\alpha))^2 d\alpha$$

$$= 2\int_0^1 \alpha a_1(\alpha)^2 d\alpha - E_L(Tac(Q,S))^2$$

$$= 2\int_0^1 \alpha \left(\frac{(c_{01} - a_{01}\sqrt{\frac{1}{\alpha}-1})D}{Q} + \frac{(c_{h1} - a_{h1}\sqrt{\frac{1}{\alpha}-1})(Q-S)^2}{2Q} + \frac{(c_{s1} - a_{s1}\sqrt{\frac{1}{\alpha}-1})(S)^2}{2Q} \right)^2 d\alpha$$

$$- \left(\frac{(8c_{01} - 2a_{01}\pi)D + (4c_{h1} - a_{h1}\pi)(Q-S)^2 + (4c_{s1} - a_{s1}\pi)S^2}{8Q} \right)^2$$

$$= \frac{1}{64Q^2}\left[64(a_{01})^2) - 4(a_{01})^2\pi^2\}D^2 + \{16(a_{h1})^2) - (a_{h1})^2\pi^2\}(Q-S)^4 + \{16(a_{s1})^2) - (a_{s1})^2\pi^2\}S^4 + \{64a_{01}a_{h1} - \right.$$

$$\left. 4a_{01}a_{h1}\pi^2\}D(Q-S)^2 + \{32a_{h1}a_{s1} - 2a_{h1}a_{s1}\pi^2\}(Q-S)^2S^2 + \{64a_{01}a_{s1} - 4a_{01}a_{s1}\pi^2\}DS^2\right]. \quad (9.4.26)$$

$$Var_R(Tac(Q,S)) = 2\int_0^1 \alpha(E_R(Tac(Q,S)) - a_2(\alpha))^2 d\alpha$$

$$= 2\int_0^1 \alpha a_2(\alpha)^2 d\alpha - E_R(Tac(Q,S))^2$$

$$= 2\int_0^1 \alpha \left(\frac{(c_{01} + a_{01}\sqrt{\frac{1}{\alpha}-1})D}{Q} + \frac{(c_{h1} + a_{h1}\sqrt{\frac{1}{\alpha}-1})(Q-S)^2}{2Q} + \frac{(c_{s1} + a_{s1}\sqrt{\frac{1}{\alpha}-1})(S)^2}{2Q} \right)^2 d\alpha$$

$$- \left(\frac{(8c_{01} + 2a_{01}\pi)D + (4c_{h1} + a_{h1}\pi)(Q-S)^2 + (4c_{s1} + a_{s1}\pi)S^2}{8Q} \right)^2$$

$$= \frac{1}{64Q^2}\left[64(a_{01})^2) - 4(a_{01})^2\pi^2\}D^2 + \{16(a_{h1})^2) - (a_{h1})^2\pi^2\}(Q-S)^4 + \{16(a_{s1})^2) - (a_{s1})^2\pi^2\}S^4 + \{64a_{01}a_{h1} - \right.$$

$$\left. 4a_{01}a_{h1}\pi^2\}D(Q-S)^2 + \{32a_{h1}a_{s1} - 2a_{h1}a_{s1}\pi^2\}(Q-S)^2S^2 + \{64a_{01}a_{s1} - 4a_{01}a_{s1}\pi^2\}DS^2\right]. \quad (9.4.27)$$

So $Var_L(Tac(Q,S)) = Var_R(Tac(Q,S))$

Now $Var(Tac(Q,S)) = \frac{Var_L(Tac(Q,S)) + Var_R(Tac(Q,S))}{2}$

$$= \frac{2Var_L(Tac(Q,S))}{2}, \qquad [Var_L(Tac(Q,S)) = Var_R(Tac(Q,S))].$$

$$= Var_L(Tac(Q.S)) \qquad\qquad (9.4.28)$$

i.e., $Var_L(Tac(Q,S)) = Var_R(Tac(Q,S)) = Var(Tac(Q,S))$

$$= \frac{1}{64Q^2}[64(a_{01})^2) - 4(a_{01})^2\pi^2\}D^2 + \{16(a_{h1})^2) - (a_{h1})^2\pi^2\}(Q-S)^4 + \{16(a_{s1})^2) - (a_{s1})^2\pi^2\}S^4 + \{64a_{01}a_{h1} - 4a_{01}a_{h1}\pi^2\}D(Q-S)^2 + \{32a_{h1}a_{s1} - 2a_{h1}a_{s1}\pi^2\}(Q-S)^2S^2 + \{64a_{01}a_{s1} - 4a_{01}a_{s1}\pi^2\}DS^2]. \qquad (9.4.29)$$

Theorem 9.4.2.6

The standard deviation $Sd(Tac(Q.S)$ defined as follows;

$$Sd(Tac(Q,S)) = \frac{1}{64Q^2}[64(a_{01})^2) - 4(a_{01})^2\pi^2\}D^2 + \{16(a_{h1})^2) - (a_{h1})^2\pi^2\}(Q-S)^4 + \{16(a_{s1})^2) - (a_{s1})^2\pi^2\}S^4 +$$
$$\{64a_{01}a_{h1} - 4a_{01}a_{h1}\pi^2\}D(Q-S)^2 + \{32a_{h1}a_{s1} - 2a_{h1}a_{s1}\pi^2\}(Q-S)^2S^2 + \{64a_{01}a_{s1} - 4a_{01}a_{s1}\pi^2\}DS^2]^{0.5} \quad (9.4.30)$$

Proof

We know $Sd(Tac(Q,S)) = +\sqrt{Var(Tac(Q,S))}$

But

$$Var(Tac(Q,S)) = \frac{1}{64Q^2}[64(a_{01})^2) - 4(a_{01})^2\pi^2\}D^2 + \{16(a_{h1})^2) - (a_{h1})^2\pi^2\}(Q-S)^4 + \{16(a_{s1})^2) - (a_{s1})^2\pi^2\}S^4 + \{64a_{01}a_{h1} - 4a_{01}a_{h1}\pi^2\}D(Q-S)^2 + \{32a_{h1}a_{s1} - 2a_{h1}a_{s1}\pi^2\}(Q-S)^2S^2 + \{64a_{01}a_{s1} - 4a_{01}a_{s1}\pi^2\}DS^2],$$

i.e.,

$$Sd(Tac(Q,S)) = \frac{1}{64Q^2}[64(a_{01})^2) - 4(a_{01})^2\pi^2\}D^2 + \{16(a_{h1})^2) - (a_{h1})^2\pi^2\}(Q-S)^4 + \{16(a_{s1})^2) - (a_{s1})^2\pi^2\}S^4 +$$
$$\{64a_{01}a_{h1} - 4a_{01}a_{h1}\pi^2\}D(Q-S)^2 + \{32a_{h1}a_{s1} - 2a_{h1}a_{s1}\pi^2\}(Q-S)^2S^2 + \{64a_{01}a_{s1} - 4a_{01}a_{s1}\pi^2\}DS^2]^{0.5}$$

Theorem 9.4.2.7

Crisp possibilistic function $E(Tac(Q,S))$ is a convex function.

Proof: From (9.4.18) we have,

$$\frac{\partial E(Tac(Q,S))}{\partial Q} = \frac{-c_{01}D}{Q^2} + \frac{c_{h1}}{2}\left(1 - \frac{S^2}{Q^2}\right) - \frac{c_{s1}}{2}\frac{S^2}{Q^2},$$

$$\frac{\partial E(Tac(Q,S))}{\partial S} = \frac{c_{h1}}{2}(-Q + 2S) + \frac{c_{s1}}{2}\frac{S}{Q}.$$

And

$$\frac{\partial^2 E(Tac(Q,S))}{\partial Q^2} = \frac{2c_{01}D + c_{h1}S^2 + c_{s1}S^2}{Q^3} > 0,$$

$$\frac{\partial^2 E(\text{Tac}(Q,S))}{\partial S^2} = \frac{c_{h1}+c_{S1}}{Q} > 0.$$

Here the Hessian determinant of order 2 is given by

$$D_2 = \begin{vmatrix} C_{11} & C_{12} \\ C_{21} & C_{22} \end{vmatrix}, \quad C_{ij} = \frac{\partial^2 Tac_1(t_1{}^*,t_2{}^*)}{\partial t_i \partial t_j} \quad (i,j=1,2).$$

$$= \begin{vmatrix} \frac{2c_{01}D+c_{h1}S^2+c_{s1}S^2}{Q^3} & -S(\frac{c_{h1}+c_{s1}}{Q^2}) \\ -S(\frac{c_{h1}+c_{s1}}{Q^2}) & \frac{c_{h1}+c_{s1}}{Q} \end{vmatrix}$$

$$= \frac{2c_{01}D+c_{h1}S^2+c_{s1}S^2}{Q^3} \times (\frac{c_{h1}+c_{s1}}{Q}) - S^2(\frac{c_{h1}+c_{s1}}{Q^2})^2$$

$$= (\frac{c_{h1}+c_{s1}}{Q^4})(2c_{01}D + c_{h1}S^2 + c_{s1}S^2 - S^2c_{h1} - S^2c_{s1})$$

$$= 2c_{01}D(\frac{c_{h1}+c_{s1}}{Q^4}) > 0. \tag{9.4.31}$$

It seen that that $D_2 > 0$, also $D_1 = C_{11} = \frac{2c_{01}D+c_{h1}S^2+c_{s1}S^2}{Q^3} > 0.$

i.e., $E(\text{Tac}(Q,S))$ is convex.

9.5   Numerical Solution

9.5.1   General case:

Let the input value of D = 10 and $\mu_{c_0}(x)$, $\mu_{c_h}(x)$, $\mu_{c_s}(x)$ be fuzzy membership function of fuzzy numbers c_0, c_h and c_s respectively. Where,

$$\mu_{c_0}(x) = \begin{cases} 0 & x \leq 80 \\ \dfrac{x-80}{90-80} & 80 \leq x \leq 90 \\ 1 & 90 \leq x \leq 110, \\ \dfrac{x-120}{110-120} & 110 \leq x \leq 120 \\ 0 & x \geq 120 \end{cases}$$

$$\mu_{c_h}(x) = \begin{cases} 0 & x \leq 120 \\ \dfrac{x-120}{130-120} & 120 \leq x \leq 130 \\ 1 & 130 \leq x \leq 150, \\ \dfrac{x-160}{150-160} & 150 \leq x \leq 160 \\ 0 & x \geq 160 \end{cases}$$

$$\mu_{c_s}(x) = \begin{cases} 0 & x \leq 60 \\ \dfrac{x-60}{70-60} & 60 \leq x \leq 70 \\ 1 & 70 \leq x \leq 90, \\ \dfrac{x-100}{90-100} & 90 \leq x \leq 100 \\ 0 & x \geq 100 \end{cases}$$

Therefore,

$$C_0(\alpha) = [C_{01}(\alpha), C_{02}(\alpha)] = [80 + \alpha(90 - 80), 120 + \alpha(110 - 120)],$$

$$C_h(\alpha) = [C_{h1}(\alpha), C_{h2}(\alpha)] = [120 + \alpha(130 - 120), 160 + \alpha(150 - 160)],$$

$$C_s(\alpha) = [C_{s1}(\alpha), C_{s2}(\alpha)] = [60 + \alpha(70 - 60), 100 + \alpha(90 - 100)].$$

Then optimal Q, S and Tac(Q,S) in possibilistic setup from (9.4.7) to (5.4.15) are follows,

Table-9.1(Out-put vales of lower possibilistic mean for general case)

$E_L(Q)$	$E_L(S)$	$E_L(Tac(Q,S))$
6.300	4.127	275.152

Table-9.2(Out-put vales of upper possibilistic mean for general case)

$E_R(Q)$	$E_R(S)$	$E_R(Tac(Q,S))$
6.250	3.885	362.640

Table-9.3 (Out-put vales of possibilistic mean for general case)

$E(Q)$	$E(S)$	$E(Tac(Q,S))$
6.268	3.989	319.090

Here IVPM (Tac(Q,S)) = [275.152, 373.489]

9.5.1.1 Sensitivity Analysis

We now examine to sensitivity analysis of the optimal solution of the inventory model for changes of D, keeping the other parameters unchanged.

Table-9.4 (Sensitivity analysis for general case)

D	% of change	$E_L(Q)$	$E(Q)$	$E_R(Q)$	$E_L(S)$	$E(S)$	$E_R(S)$	$E_L(Tac(Q,S)$	$E(Tac(Q,S))$	$E_R(Tac(Q,S)$
2	-80	2.817	2.803	2.795	1.846	1.784	1.738	123.095	144.457	162.950

4	-60	3.984	3.964	3.953	2.610	2.523	2.457	153.408	174.365	234.762
6	-40	4.880	4.855	4.842	3.197	3.089	3.010	283.062	205.273	281.256
8	-20	5.634	5.606	5.591	3.691	3.567	3.475	213.716	235.181	330.321
10	0	6.300	6.268	6.250	4.127	3.989	3.885	243.152	355.090	362.640
12	+20	6.900	6.866	6.847	4.521	4.369	4.256	373.023	376.998	390.432
14	+40	7.453	7.416	7.396	4.883	4.719	4.597	303.676	386.906	420.321
16	+60	7.968	7.928	7.906	5.221	5.045	4.915	335.330	404.814	450.654
18	+80	8.451	8.409	8.386	5.537	5.351	5.213	369.162	428.722	486.582

Effect, for increment parameters-

From the table (Table-9.4) we see that all of decision variables increase for increasing of D.

9.5.2 Particular case:

Let crisp input value of D = 10 and $\mu_{c_0}(x)$, $\mu_{c_h}(x)$, $\mu_{c_s}(x)$ be Bell fuzzy membership function of fuzzy numbers c_0, c_h and c_s respectively. Where,

$$\mu_{c_0}(x) = \frac{1}{1+\left|\frac{x-100}{50}\right|^2}, \ \mu_{c_h}(x) = \frac{1}{1+\left|\frac{x-200}{100}\right|^2}, \ \mu_{c_s}(x) = \frac{1}{1+\left|\frac{x-50}{30}\right|^2}.$$

So $\quad C_0(\alpha) = [C_{01}(\alpha), C_{02}(\alpha)] = \left[100 - 50 \times \sqrt{\frac{1}{\alpha}-1}, 100 + 50 \times \sqrt{\frac{1}{\alpha}-1}\right],$

$\quad C_h(\alpha) = [C_{h1}(\alpha), C_{h2}(\alpha)] = \left[200 - 100 \times \sqrt{\frac{1}{\alpha}-1}, 200 + 100 \times \sqrt{\frac{1}{\alpha}-1}\right],$

$\quad C_s(\alpha) = [C_{s1}(\alpha), C_{s2}(\alpha)] = \left[50 - 30 \times \sqrt{\frac{1}{\alpha}-1}, 50 + 30 \times \sqrt{\frac{1}{\alpha}-1}\right].$

Then optimal Q, S and Tac(Q,S) in possibilistic setup from (9.4.16) to (9.4.25) are follows,

Table-9.5 (Out-put vales of lower possibilistic mean for particular case)

$E_L(Q)$	$E_L(S)$	$E_L(Tac(Q,S))$
7.480	6.143	162.344

Table-9.6 (Out-put vales of upper possibilistic mean for particular case)

$E_R(Q)$	$E_R(S)$	$E_R(Tac(Q,S))$
6.918	5.473	402.654

Table-9.7 (Out-put vales of possibilistic mean for particular case)

$E(Q)$	$E(S)$	$E(Tac(Q,S))$
7.071	5.657	282.449

Here IVPM (Tac(Q,S)) = [162.344, 402.654].

9.5.2.1 Sensitivity Analysis

We now examine to sensitivity analysis of the optimal solution of the inventory model for changes of D, keeping the other parameters unchanged.

Table-9.8(Sensitivity analysis for particular case)

D	% of change	$E_L(Q)$	$E(Q)$	$E_R(Q)$	$E_L(S)$	$E(S)$	$E_R(S)$	$E_L(Tac(Q,S))$	$E(Tac(Q,S))$	$E_R(Tac(Q,S))$
2	-80	3.345	3.162	3.094	2.747	2.512	2.447	72.604	125.339	180.074
4	-60	4.731	4.472	4.376	3.885	3.645	3.461	102.703	182.001	250.321
6	-40	5.794	5.477	5.359	4.758	4.483	4.240	125.802	218.062	310.432
8	-20	6.690	6.324	6.188	5.494	5.123	4.895	146.067	253.075	360.083
10	0	7.480	7.071	6.918	6.143	5.657	5.473	162.344	282.449	402.654
12	+20	8.194	7.746	7.579	6.729	6.232	5.995	177.578	310.876	442.654
14	+40	8.850	8.367	8.186	7.268	6.751	6.476	191.812	339.752	478.732
16	+60	9.461	8.944	8.751	7.770	7.325	6.923	204.045	353.399	502.654
18	+80	10.03	9.487	9.282	8.241	7.802	7.343	218.363	379.510	540.217

Effect, for increment parameters-

From the table (Table-9.4) we see that all of decision variables increase for increasing of D.

9.6   Conclusion

In this chapter we have developed a Bell shaped fuzzy approach to the EOQ model. The model is developed with shortage under fully backlogged. Here we have discussed only a particular Bell shape membership function (MF) (taking b = 1). In future general form of Bell shape MF would be used. The methodology proposed in this chapter may also be applicable to other economic order quantity (EOQ) or economic production quality (EPQ) models. Our approach provide here a simple EOQ model, but in future it should be used many complex EOQ or EPQ model. For future research of uncertainty in inventory model, by using different type of fuzzy numbers such as random or adaptive fuzzy demand rate or generalize fuzzy number be analytically more challenging and interesting.

Chapter 10

Fuzzy EOQ Model with Constant Demand and Shortages: A Fuzzy Signomial Geometric Programming (FSGP) Approach

Richard J. Duffin and Elmor L. Peterson introduced the term "signomial" in their seminal joint work on general algebraic optimization, published in the late 1960s and early 1970s. A fuzzy signomial geometric programming (FSGP) optimization problem often provides a much more accurate mathematical representation of real-world nonlinear optimization problems.

In this chapter, a fuzzy economic order quantity (EOQ) model with shortages under fully backlogging and constant demand is formulated and solved by fuzzy signomial geometric programming (FSGP) technique. *Fuzzy signomial geometric programming (FSGP) technique provides a powerful technique for solving some special types of non-linear problems. Here we have proposed a new idea of fuzzy modified signomial geometric programming (FMSGP) and some necessary theorems have been derived. Finally, the model is illustrated by some numerical examples and applications.*

10.1   Deterministic EOQ Model

In many real-life situations shortages occur in an EOQ model. So, we have developed an EOQ model with shortage and costs are incurred. The notations to be used are:

Tac(Q,S): Total average cost of the EOQ model.

Q: Order quantity.

$Q - S$: Maximum shortage that occurs under an ordering policy

c_s: Carrying cost per item per unit time.

c_h: Shortages cost per item per unit time.

c_0: Ordering cost per order.

D: Demand rate per unit time.

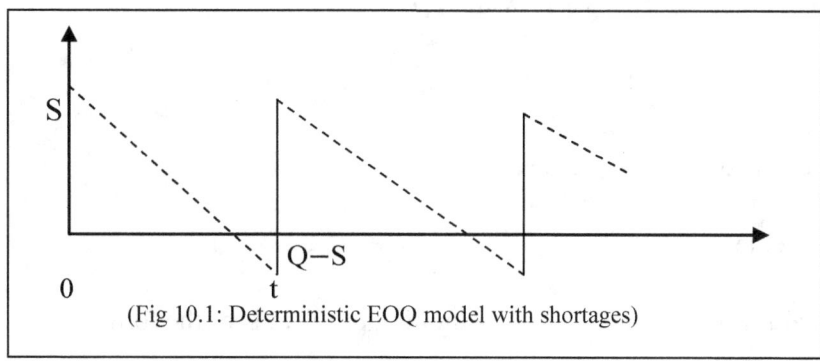

(Fig 10.1: Deterministic EOQ model with shortages)

Variables of this EOQ model are Q, S and c_0, c_h, c_s are constant parameters.

Thus,

Total carrying cost $= \frac{c_s S^2}{2D}$,

Total shortages cost $= \frac{c_h(q-s)^2}{2D}$,

So total cost $= c_0 + \frac{c_h(q-s)^2}{2D} + \frac{c_s S^2}{2D}$

And total average cost $\text{Tac(Q,S)} = \frac{1}{t}\left[c_0 + \frac{c_h(q-s)^2}{2D} + \frac{c_s S^2}{2D}\right]$

$$= \frac{c_0 D}{Q} + \frac{c_h(Q-S)^2}{2Q} + \frac{c_s(S)^2}{2Q}, \qquad \left[t = \frac{Q}{D}\right].$$

i.e., problem is

$$\text{Minimize Tac(Q,S)} = \frac{c_0 D}{Q} + \frac{c_h(Q-S)^2}{2Q} + \frac{c_s(S)^2}{2Q} \qquad (10.1.1)$$

Subject to   Q, S > 0.

10.2   Fuzzy EOQ Model

For uncertainty of cost parameters, we take the parameters $\tilde{c}_0$, $\tilde{c}_h$ and $\tilde{c}_s$ are in fuzzy numbers, then from (10.1.1) we have

$$\text{Minimize } \widetilde{Tac}(Q,S) = \frac{\tilde{c}_0 D}{Q} + \frac{\tilde{c}_h(Q-S)^2}{2Q} + \frac{\tilde{c}_s(S)^2}{2Q} \qquad (10.2.1)$$

Subject to   Q, S > 0.

10.3   Unconstrained Fuzzy Signomial Geometric Programming (UFSGP) Problem

A problem without any restrictions is called unconstrained problem. I.e., a problem of the form

Minimize $\quad g_0(x_1, x_2, \ldots\ldots\ldots\ldots, x_m)$ $\hfill$ (10.3.1)

Subject to $\quad x_j > 0,\ j = 1, 2, \ldots\ldots, m,$

is called unconstrained problem.

Primal problem:

A primal fuzzy signomial GP programming problem is of the form

Minimize $\quad \tilde{g}_0(x_1, x_2, \ldots\ldots\ldots\ldots, x_m)$ $\hfill$ (10.3.2)

Subject to $\quad x_j > 0,\ j = 1, 2, \ldots\ldots, m.$

Where $\quad \tilde{g}_0(x) = \sum_{i=1}^{k} \sigma_i \tilde{c}_i \prod_{j=1}^{m} x_j^{a_{ij}}.$

Here a_{ij} are real numbers, $\sigma_i = \pm 1$ and coefficient $\tilde{c}_i$ are fuzzy triangular, defined as $\tilde{c}_i = (c_i^1, c_i^2, c_i^3)$.

Using nearest interval approximation (NIA) method, transformed triangular fuzzy number into interval number i.e., $[c_i^L, c_i^U]$. Then the fuzzy signomial geometric programming problem is of the following form

Min $\quad \hat{g}_0(x) = \sum_{i=1}^{k} \sigma_i \hat{c}_i \prod_{j=1}^{m} x_j^{a_{ij}}$ $\hfill$ (10.3.3)

Subject to $\quad x_j > 0,\ j = 1, 2, \ldots\ldots, m.$

Where $\hat{c}_i$ denotes the interval counter parts i.e., $\hat{c}_i \in [c_i^L, c_i^U].\, c_i^L > 0, c_i^U > 0$, for all i. Using parametric interval-valued functional form, the problem (10.3.3) reduces to

Min $\quad g_0(x, s) = \sum_{i=1}^{k} \sigma_i (c_i^L)^{1-s} (c_i^U)^s \prod_{j=1}^{m} x_j^{a_{ij}}$ $\hfill$ (10.3.4)

Subject to $\quad x_j > 0,\ j = 1, 2, \ldots\ldots, m.$

This is a parametric signomial geometric programming (PSGP) problem.

Dual signomial GP problem:

Dual GP problem of the given primal GP problem is

Maximize $v(\delta, s) = \zeta_0 \left[\prod_{i=1}^{n} \left(\frac{(c_i^L)^{1-s}(c_i^U)^s}{\delta_i} \right)^{\sigma_i \delta_i} \right]^{\zeta_0}$ $\hfill$ (10.3.5)

Subject to

$$\sum_{i=1}^{k} \sigma_i \delta_i = \zeta_0,$$

$$\sum_{i=1}^{k} \sigma_i a_{ij} \delta_i = 0, \qquad\qquad j = 1,2,\ldots\ldots\ldots,m$$

$$\delta_i > 0.$$

Case I: n > m+1, (i.e. DD > 0) so the DP presents a system of linear equations for the dual variables. Here the number of linear equations is less than the number of dual variables. More solutions of dual variable vector exist. In order to find an optimal solution of DP, we need to use some algorithmic methods.

Case II: n < m+1, (i.e. DD < 0) so the DP presents a system of linear equations for the dual variables. Here the number of linear equations is greater than the number of dual variables. In this case generally no solution vector exists for the dual variables. However, using Least Square (LS) or Min-Max (MM) method one can get an approximate solution for this system.

From the primal-dual variable relationship, we get is

$$(c_i{}^L)^{1-s}(c_i{}^U)^s \prod_{j=1}^{m} x_j{}^{*a_{ij}} = \zeta_0 \delta^*{}_i \, v(\delta^*,s^*). \tag{10.3.6}$$

Note 10.3.1: A Weak Duality theorem would say that

$$g_0(x,s) \geq v(\delta,s)$$

For any primal-feasible x and dual-feasible δ but this is not true of the pseudo-dual fuzzy signomial GP problem.

10.3.2 Corollary: When the value of σ_i is 1, then a fuzzy signomial geometric programming (FSGP) problem transform to ordinary geometric programming problem.

Theorem 10.3.3: When σ_i is 1, then $g_o(x, s) \geq v(\delta,s)$ (Primal-Dual Inequality).

Proof

The expression for $g_0(x, s)$ can be written as

$$g_o(x, s) = \sum_{i=1}^{n} \delta_k \left(\frac{(c_i{}^L)^{1-s}(c_i{}^U)^s \prod_{j=1}^{m} x_j{}^{\alpha_{kj}}}{\delta_k} \right).$$

Here the weights are $\delta_1, \delta_2, \dots\dots\dots, \delta_n$ and positive terms are $\dfrac{(c_1{}^L)^{1-s}(c_1{}^U)^s \prod_{j=1}^m x_j{}^{\alpha_{1j}}}{\delta_1}$,

$\dfrac{(c_2{}^L)^{1-s}(c_2{}^U)^s \prod_{j=1}^m x_j{}^{\alpha_{2j}}}{\delta_2}, \dots\dots\dots, \dfrac{(c_n{}^L)^{1-s}(c_n{}^U)^s \prod_{j=1}^m x_j{}^{\alpha_{nj}}}{\delta_n}$.

Now applying AM.-.GM inequality, we get

$$\left(\frac{(c_1{}^L)^{1-s}(c_1{}^U)^s \prod_{j=1}^m x_j{}^{\alpha_{1j}} + (c_2{}^L)^{1-s}(c_2{}^U)^s \prod_{j=1}^m x_j{}^{\alpha_{2j}} + \dots + (c_n{}^L)^{1-s}(c_n{}^U)^s \prod_{j=1}^m x_j{}^{\alpha_{nj}}}{(\delta_1 + \delta_2 + \dots + \delta_n)}\right)^{(\delta_{01}+\delta_{02}+\dots+\delta_n)}$$

$$\geq \left(\left(\frac{(c_1{}^L)^{1-s}(c_n{}^U)^s \prod_{j=1}^m x_j{}^{\alpha_{1j}}}{\delta_1}\right)^{\delta_1}\left(\frac{(c_2{}^L)^{1-s}(c_2{}^U)^s \prod_{j=1}^m x_j{}^{\alpha_{2j}}}{\delta_2}\right)^{\delta_2}\dots\left(\frac{(c_n{}^L)^{1-s}(c_n{}^U)^s \prod_{j=1}^m x_j{}^{\alpha_{nj}}}{\delta_n}\right)^{\delta_n}\right)$$

Or $\qquad \left(\dfrac{g_0(x,s)}{\sum_{i=1}^n \delta_i}\right)^{\sum_{i=1}^n \delta_i} \geq \prod_{i=1}^n \left(\dfrac{(c_i{}^L)^{1-s}(c_i{}^U)^s \prod_{j=1}^m x_j{}^{\alpha_{nj}}}{\delta_i}\right)^{\delta_i}$ $\qquad [as \sum_{i=1}^n \delta_k = 1]$

Or $\qquad g_0(x,s) \geq \left(\dfrac{(c_i{}^L)^{1-s}(c_i{}^U)^s}{\delta_k}\right)^{\sum_{i=1}^n \delta_i} \prod_{j=1}^m x_j{}^{\sum_{i=1}^n \alpha_{ij}\delta_i}$

Or $\qquad g_0(x,s) \geq \prod_{i=1}^n \left(\dfrac{(c_i{}^L)^{1-s}(c_i{}^U)^s}{\delta_i}\right)^{\delta_i} \prod_{j=1}^m x_j{}^{\sum_{i=1}^n \alpha_{ij}\delta_i}$

$$= \prod_{i=1}^n \left(\frac{(c_i{}^L)^{1-s}(c_i{}^U)^s}{\delta_i}\right)^{\delta_i} = v(\delta,s) \qquad [as \sum_{k=1}^{T_0} \alpha_{0kj}\delta_{ok} = 0]$$

i.e., $\qquad g_0(x, s) \geq v(\delta,s)$.

Ex. 10.3.4: Minimize $\quad \widetilde{Tac}(Q,S) = \dfrac{\widetilde{c_0}D}{Q} + \dfrac{\widetilde{c_h}(Q-S)^2}{2Q} + \dfrac{\widetilde{c_s}(S)^2}{2Q}$

$\qquad\qquad$ Subject to $\quad Q, S > 0$.

With input values

Table-10.1 (Fuzzy input data for FSGP problem)

$\widetilde{c_0} = \widetilde{20}$	$\widetilde{c_h} = \widetilde{50}$	$\widetilde{c_s} = \widetilde{50}$	D
(16, 20, 24)	(40, 50, 60)	(40, 50, 60)	10

Using nearest approximation (NIA) method

$\widetilde{20} = (16, 20, 24) \approx [18, 22] \approx 18^{1-s}22^s \in [18, 22]$;

$\widetilde{50} = (40, 50, 60) \approx [45, 55] \approx 45^{1-s}55^s \in [45, 55]; s \in [0, 1]$.

Then the problem reduce to

$$\text{Min. Tac(Q,S)} = \frac{18^{1-s}22^s.10}{Q} + \frac{45^{1-s}55^s(Q-S)^2}{2Q} + \frac{45^{1-s}55^s(S)^2}{2Q}$$

Sub. $Q, S > 0.$

i.e., $\text{Min. Tac(Q,S)} = \frac{18^{1-s}22^s.10}{Q} + \frac{45^{1-s}55^s(S)^2}{Q} + \frac{45^{1-s}55^sQ}{2} - \frac{45^{1-s}55^sS}{1}$

Sub. $Q, S > 0.$

This is a primal problem and corresponding dual problem is

$$v(\delta, s) = \left(\frac{10.16^{1-s}20^s}{\delta_1}\right)^{\delta_1} \left(\frac{45^{1-s}55^s}{\delta_2}\right)^{\delta_2} \left(\frac{45^{1-s}55^s}{2\delta_3}\right)^{\delta_3} \left(\frac{45^{1-s}55^s}{\delta_4}\right)^{-\delta_4}$$

Subject to

$$\delta_1 + \delta_2 + \delta_3 - \delta_4 = 1,$$

$$-\delta_1 - \delta_2 + \delta_3 = 0,$$

$$2\delta_2 - \delta_4 = 0,$$

Solving above equations, we have

$$\delta_4 = 2\delta_2,$$

$$\delta_3 = 1 - \delta_1 - \delta_2 + \delta_4 = 1 - \delta_1 - \delta_2 + 2\delta_2 = 1 - \delta_1 + \delta_2,$$

$$\delta_1 + \delta_2 = 1 - \delta_1 + \delta_2, \delta_1 = \frac{1}{2}, \delta_3 = \frac{1}{2} + \delta_2.$$

i.e., $v(\delta, s) = \left(\frac{10.16^{1-s}24^s}{1/2}\right)^{1/2} \left(\frac{45^{1-s}55^s}{\delta_2}\right)^{\delta_2} \left(\frac{45^{1-s}55^s}{2(0.5+\delta_2)}\right)^{(0.5+\delta_2)} \left(\frac{45^{1-s}55^s}{2\delta_2}\right)^{-2\delta_2}$ (10.3.7)

Taking log on both side of (10.3.7) and then partially differentiating with respect to δ_2 and using the conditions of finding optimal solution we get this equation

$$(log\, 2.45^{1-s}55^s - log2\delta_2) + (log45^{1-s}55^s - log2(0.5 + \delta_2) - 2(log45^{1-s}55^s - log2\delta_2) = 0$$

$$\Rightarrow log(2.45^{1-s}55^s)/45^{1-s}55^s - log2\delta_2(1 + 2\delta_2) = 0$$

$$\Rightarrow \delta_2(1 + 2\delta_2) = 1.$$

From primal-dual relation

$$\frac{10.16^{1-s}24^s}{Q} = \delta_1 v(\delta, s),$$

$$\frac{45^{1-s}55^s S^2}{2Q} = \delta_2 v(\delta, s),$$

$$\frac{45^{1-s}55^s Q}{2} = \delta_3 v(\delta, s),$$

$$45^{1-s}55^s S = \delta_4 v(\delta, s).$$

Solving above relations with difference values of weight, we get the list of values in table-10.2.

Table -10.2 (optimal solution for FSGP problem)

				Optimal values objectives	
s	$1-s$	Optimal dual variables	Optimal primal variables	$v(\delta, s)$	$Tac(Q, S)$
0.1	0.9	$\delta_1^* = 0.5,\ \delta_2^* = 0.5,$ $\delta_3^* = 1.0, \delta_4^* = 1.0.$	$S^* = 1.905$ $Q^* = 3.810$	87.464	87.464
0.3	0.7	$\delta_1^* = 0.5,\ \delta_2^* = 0.5,$ $\delta_3^* = 1.0, \delta_4^* = 1.0.$	$S^* = 1.944$ $Q^* = 3.889$	92.929	92.929
0.5	0.5	$\delta_1^* = 0.5,\ \delta_2^* = 0.5,$ $\delta_3^* = 1.0, \delta_4^* = 1.0.$	$S^* = 1.984$ $Q^* = 3.969$	98.736	98.736
0.7	0.3	$\delta_1^* = 0.5,\ \delta_2^* = 0.5,$ $\delta_3^* = 1.0, \delta_4^* = 1.0.$	$S^* = 2.026$ $Q^* = 4.051$	104.906	104.906
0.9	0.1	$\delta_1^* = 0.5,\ \delta_2^* = 0.5,$ $\delta_3^* = 1.0, \delta_4^* = 1.0.$	$S^* = 2.068$ $Q^* = 4.135$	111.462	111.462

10.4 Unconstrained Fuzzy Modified Signomial Geometric Programming (UFMSGP) Problem

Primal problem:

A primal modified signomial GP programming problem is of the form

Minimize $\quad \tilde{g}_0(x_{lj})$ $\qquad\qquad\qquad\qquad\qquad$ (10.4.1)

Subject to $\quad x_{lj} > 0,\ j = 1, 2, \ldots\ldots, m.$

Where $\tilde{g}_0(x) = \sum_{l=1}^n \sum_{i=1}^k \sigma_{li} \tilde{c}_{li} \prod_{j=1}^m x_{lj}{}^{a_{lij}}$.

Here a_{lij} are real numbers, $\sigma_{li} = \pm 1$ and coefficient $\tilde{c}_{li}$ are fuzzy triangular, defined as $\tilde{c}_{li} = (c_{li}{}^1, c_{li}{}^2, c_{li}{}^3)$.

Using nearest interval approximation (NIA) method, triangular fuzzy number transformed into interval number i.e., $[c_{li}{}^L, c_{li}{}^U]$. Then the fuzzy modified signomial geometric programming (FMSGP) problem is of the following form

Minimize $\quad \hat{g}_0(x_1, x_2, \ldots \ldots \ldots, x_m)$ $\qquad\qquad\qquad$ (10.4.2)

Subject to $\quad x_j > 0,\ j = 1, 2, \ldots \ldots, m.$

Where $\hat{g}_0(x) = \sum_{l=1}^n \sum_{i=1}^k \sigma_{li} \hat{c}_{li} \prod_{j=1}^m x_j{}^{a_{lij}}$.

Where $\hat{c}_{li}$ denotes the interval counter parts i.e., $\hat{c}_{li} \in [c_{li}{}^L, c_{li}{}^U] . c_{li}{}^L > 0, c_{li}{}^U > 0$, for all i. Using parametric interval-valued functional form, the problem (10.4.2) reduces to

Minimize $\quad g_0(x_1, x_2, \ldots \ldots \ldots, x_m, s)$ $\qquad\qquad\qquad$ (10.4.3)

Subject to $\quad x_j > 0,\ j = 1, 2, \ldots \ldots, m.$

Where $g_0(x, s) = \sum_{l=1}^n \sum_{i=1}^k \sigma_{li} (c_{li}{}^L)^{1-s} (c_{li}{}^U)^s \prod_{j=1}^m x_j{}^{a_{lij}}$.

This is a parametric modified signomial geometric programming (PMSGP) problem.

Dual modified signomial GP problem:

Dual GP problem of the given primal GP problem is

Maximize $v(\delta, s) = \zeta_0 \left[\prod_{l=1}^n \prod_{i=1}^k (\frac{(c_{li}{}^L)^{1-s}(c_{li}{}^U)^s}{\delta_{li}})^{\sigma_{li}\delta_{li}} \right]^{\zeta_0}$ $\qquad\qquad$ (10.4.4)

Subject to $\sum_{l=1}^n \sum_{i=1}^k \sigma_{li}\delta_{li} = \zeta_0,$

$\qquad\qquad \sum_{l=1}^n \sum_{i=1}^k \sigma_{li} a_{lij} \delta_{li} = 0, \qquad\qquad j = 1,2,\ldots\ldots\ldots, m.$

$\qquad\qquad \delta_{li} > 0,$

Case I: nk $\geq$ nm+n, (i.e. DD > 0), so the DP presents a system of linear equations for the dual variables. Here the number of linear equations is less than the number of dual variables. More solutions of dual variable vector exist. In order to find an optimal solution of DP, we need to use some algorithmic methods.

Case II: nk < nm+n, (i.e. DD < 0), so the DP presents a system of linear equations for the dual variables. Here the number of linear equations is greater than the number of dual variables. In this case generally no solution vector exists for the dual variables. However, using Least Square (LS) or Min-Max (MM) method one can get an approximate solution for this system.

From the primal-dual variable relationship, we have

$$(c_{li}{}^L)^{1-s}(c_{li}{}^U)^s \prod_{j=1}^m x_{lj}{}^{a_{lij}} = \zeta_0\delta^*{}_{li}\sqrt[n]{v(\delta^*,s)}., (l=1,2,\ldots\ldots,k; i=1,2,\ldots\ldots,n), s\in[0,1]. \quad (10.4.5)$$

Note 10.4.1: A Weak Duality theorem would say that

$$g_0(x_{lj},s) \geq n\sqrt[n]{v(\delta,s)}.$$

For any primal-feasible x and dual-feasible δ but this is not true of the pseudo-dual fuzzy modified signomial GP problem.

Corollary 10.4.2: When the values of σ_{li} is 1, then a fuzzy modified signomial geometric programming (FMSGP) problem transform to ordinary modified geometric programming problem.

Theorem 10.4.3: When σ_i is 1, then $g_0(x_{ij},s)\geq n\sqrt[n]{v(\delta,s)}$ (Primal- Dual Inequality).

Proof

The expression for $g_0(x_{ij},s)$ can be written as

$$g_0(x_{ij},s) =\sum_{l=1}^n \sum_{i=1}^k \delta_{ik}\left(\frac{(c_{li}{}^L)^{1-s}(c_{li}{}^U)^s\prod_{j=1}^m x_{ij}{}^{\alpha_{lij}}}{\delta_{ik}}\right).$$

Here the weights are $\delta_{l1},\delta_{l2},\ldots\ldots,\delta_{lk}$ and positive terms are $\frac{(c_{l1}{}^L)^{1-s}(c_{l1}{}^U)^s\prod_{j=1}^m x_j{}^{\alpha_{l1j}}}{\delta_{l1}},$

$$\frac{(c_{l2}{}^L)^{1-s}(c_{li}{}^U)^s\prod_{j=1}^m x_j{}^{\alpha_{l2j}}}{\delta_{l2}},\ldots\ldots,\frac{(c_{li}{}^L)^{1-s}(c_{li}{}^U)^s\prod_{j=1}^m x_j{}^{\alpha_{lnj}}}{\delta_{lk}}.$$

Now applying A.M.-.G.M inequality, we get

$$\left(\frac{\sum_{l=1}^n((c_{l1}{}^L)^{1-s}(c_{l1}{}^U)^s\prod_{j=1}^m x_{ij}{}^{\alpha_{l1j}}+(c_{l2}{}^L)^{1-s}(c_{l2}{}^U)^s\prod_{j=1}^m x_{ij}{}^{\alpha_{l2j}}+\ldots+(c_{li}{}^L)^{1-s}(c_{li}{}^U)^s\prod_{j=1}^m x_{ij}{}^{\alpha_{lkj}})}{\sum_{l=1}^n(\delta_{l1}+\delta_{l2}+\cdots+\delta_{lk})}\right)^{\sum_{l=1}^n(\delta_{l1}+\delta_{l2}+\cdots+\delta_{lk})}$$

$$\geq \sum_{l=1}^{n} \left(\left(\frac{(c_{l1}{}^{L})^{1-s}(c_{l1}{}^{U})^{s} \prod_{j=1}^{m} x_{ij}{}^{\alpha_{i1j}}}{\delta_{l1}} \right)^{\delta_{l1}} \left(\frac{(c_{l2}{}^{L})^{1-s}(c_{l2}{}^{U})^{s} \prod_{j=1}^{m} x_{ij}{}^{\alpha_{l2j}}}{\delta_{l2}} \right)^{\delta_{l2}} \cdots \left(\frac{(c_{li}{}^{L})^{1-s}(c_{li}{}^{U})^{s} \prod_{j=1}^{m} x_{ij}{}^{\alpha_{lkj}}}{\delta_{lk}} \right)^{\delta_{lk}} \right)$$

Or
$$\left(\frac{g_0(x_{ij},s)}{\sum_{l=1}^{n} \sum_{i=1}^{k} \delta_{li}} \right)^{\sum_{l=1}^{n} \sum_{i=1}^{k} \delta_{li}} \geq \prod_{l=1}^{n} \prod_{i=1}^{k} \left(\frac{(c_{li}{}^{L})^{1-s}(c_{li}{}^{U})^{s} \prod_{j=1}^{m} x_{ij}{}^{\alpha_{lij}}}{\delta_{ik}} \right)^{\delta_{ik}}$$

Or
$$\left(\frac{g_0(x_{ij},s)}{n} \right)^{n} \geq \prod_{l=1}^{n} \left(\frac{(c_{li}{}^{L})^{1-s}(c_{li}{}^{U})^{s}}{\delta_{lk}} \right)^{\sum_{l=1}^{k} \delta_{lk}} \prod_{j=1}^{m} x_{ij}{}^{\sum_{i=1}^{k} \alpha_{lij} \delta_{li}} \qquad [as \ \sum_{i=1}^{k} \delta_{li} = 1]$$

$$= \prod_{l=1}^{n} \prod_{i=1}^{k} \left(\frac{(c_{li}{}^{L})^{1-s}(c_{li}{}^{U})^{s}}{\delta_{li}} \right)^{\delta_{li}} \prod_{j=1}^{m} x_{ij}{}^{\sum_{i=1}^{k} \alpha_{lij} \delta_{li}}$$

Or
$$\left(\frac{g_0(x_{ij},s)}{n} \right)^{n} \geq \prod_{l=1}^{n} \prod_{i=1}^{k} \left(\frac{(c_{li}{}^{L})^{1-s}(c_{li}{}^{U})^{s}}{\delta_{li}} \right)^{\delta_{li}} \qquad [as \ \sum_{i=1}^{k} \alpha_{lij} \delta_{li} = 0]$$

$$= v(\delta, s)$$

i.e., $\quad g_0(x_{ij}, s) \geq n \sqrt[n]{v(\delta, s)}$.

Ex.10.4.4: Minimize $\quad \widetilde{Tac}(Q_i, S_i) = \sum_{i=1}^{n} \frac{\tilde{c}_{0i} D_i}{Q_i} + \frac{\tilde{c}_{hi}(Q_i - S_i)^2}{2Q_i} + \frac{\tilde{c}_{si}(S_i)^2}{2Q_i}$

Subject to $\quad Q_i, S_i > 0$.

With input values

Table-10.3 (Fuzzy input data for FMSGP problem)

i	$\widetilde{c_{0i}} = \widetilde{20}$	$\widetilde{c_{hi}} = \widetilde{50}$	$\widetilde{c_{si}} = \widetilde{50}$	D_i
i=1	(16, 20, 24)	(40, 50, 60)	(40, 50, 60)	10
i=2	(6, 10, 14)	(105, 125, 145)	(21, 25, 29)	15

Using nearest approximation (NIA) method

$\widetilde{10} = (6, 10, 14) \approx [8, 12] \approx 8^{1-s} 12^{s} \in [8, 12];$

$\widetilde{20} = (16, 20, 24) \approx [18, 22] \approx 18^{1-s} 22^{s} \in [18, 22];$

$\widetilde{25} = (21, 25, 29) \approx [23, 27] \approx 23^{1-s} 27^{s} \in [23, 27];$

$$\widetilde{50} = (40, 50, 60) \approx [45, 55] \approx 45^{1-s} 55^s \in [45, 55];$$

$$\widetilde{125} = (105, 125, 145) \approx [115, 135] \approx 115^{1-s} 135^s \in [115, 135]; \quad s \in [0, 1].$$

Then the problem is

$$\text{Min. Tac(Q,S,s)} = \frac{18^{1-s} 22^s . 10}{Q_1} + \frac{45^{1-s} 55^s (Q_1 - S_1)^2}{2Q_1} + \frac{45^{1-s} 55^s (S_1)^2}{2Q_1} + \frac{18^{1-s} 22^s . 10}{Q_2} + \frac{45^{1-s} 55^s (Q_2 - S_2)^2}{2Q_2} +$$

$$\frac{45^{1-s} 55^s (S_2)^2}{2Q_2}$$

Sub. $Q_1, Q_2, S_1, S_2 > 0.$

i.e.,

$$\text{Min .Tac(Q,S,s)} = \frac{18^{1-s} 22^s . 10}{Q_1} + \frac{45^{1-s} 55^s (S_1)^2}{Q_1} + \frac{45^{1-s} 55^s Q_1}{2} - \frac{45^{1-s} 55^s S_1}{1} + \frac{8^{1-s} 12^s . 15}{Q_2}$$

$$+ \frac{(115^{1-s} 135^s + 23^{1-s} 27^s)(S_2)^2}{2Q_2} + \frac{115^{1-s} 135^s Q_2}{2} - \frac{115^{1-s} 135^s S_2}{1}$$

Sub. $Q_1, Q_2, S_1, S_2 > 0.$

This is the primal problem and corresponding dual problem is

$$v(\delta, s) = \left(\frac{10.16^{1-s} 20^s}{\delta_{11}}\right)^{\delta_{11}} \left(\frac{45^{1-s} 55^s}{\delta_{12}}\right)^{\delta_{12}} \left(\frac{45^{1-s} 55^s}{2\delta_{13}}\right)^{\delta_{13}} \left(\frac{45^{1-s} 55^s}{\delta_{14}}\right)^{-\delta_{14}} \left(\frac{15.8^{1-s} 12^s}{\delta_{21}}\right)^{\delta_{21}}$$

$$\times \left(\frac{(115^{1-s} 135^s + 23^{1-s} 27^s)}{2\delta_{22}}\right)^{\delta_{22}} \left(\frac{115^{1-s} 135^s}{2\delta_{23}}\right)^{\delta_{23}} \left(\frac{115^{1-s} 135^s}{\delta_{24}}\right)^{-\delta_{24}}$$

Subject to

$$\delta_{11} + \delta_{12} + \delta_{13} - \delta_{14} = 1,$$

$$\delta_{21} + \delta_{22} + \delta_{23} - \delta_{24} = 1,$$

$$-\delta_{11} - \delta_{12} + \delta_{13} = 0,$$

$$-\delta_{21} - \delta_{22} + \delta_{23} = 0,$$

$$2\delta_{12} - \delta_{14} = 0,$$

$$2\delta_{22} - \delta_{24} = 0.$$

Solving above equations, we have

$$\delta_{14} = 2\delta_{12},$$

$\delta_{13} = 1 - \delta_{11} - \delta_{12} + \delta_{14} = 1 - \delta_{11} - \delta_{12} + 2\delta_{12} = 1 - \delta_{11} + \delta_{12}, \delta_{11} + \delta_{12} = 1 - \delta_{11} + \delta_{12}, \delta_{11} = \frac{1}{2}, \delta_{13} = \frac{1}{2} + \delta_{12}.$

And

$\delta_{24} = 2\delta_{22},$

$\delta_{23} = 1 - \delta_{21} - \delta_{22} + \delta_{24} = 1 - \delta_{21} - \delta_{22} + 2\delta_{22} = 1 - \delta_{21} + \delta_{22}, \delta_{21} + \delta_{22} = 1 - \delta_{21} + \delta_{22}, \delta_{21} = \frac{1}{2}, \delta_{23} = \frac{1}{2} + \delta_{22}.$

i.e., $v(\delta, s) = \left(\frac{10.16^{1-s}24^s}{1/2}\right)^{1/2} \left(\frac{45^{1-s}55^s}{\delta_2}\right)^{\delta_2} \left(\frac{45^{1-s}55^s}{2(0.5+\delta_2)}\right)^{(0.5+\delta_2)} \left(\frac{45^{1-s}55^s}{2\delta_2}\right)^{-2\delta_2} \left(\frac{15.8^{1-s}12^s}{1/2}\right)^{1/2}$

$\times \left(\frac{115^{1-s}135^s + 23^{1-s}27^s}{2\delta_2}\right)^{\delta_2} \left(\frac{115^{1-s}135^s}{2(0.5+\delta_2)}\right)^{(0.5+\delta_2)} \left(\frac{115^{1-s}135^s}{2\delta_2}\right)^{-2\delta_2}$ \hfill (10.4.6)

Taking log on both side of (10.5.6) and then partially differentiating with respect to δ_{12} and δ_{22} respectively and using the conditions of finding optimal solution we get;

$(log\, 2.45^{1-s}55^s - log2\delta_{12}) + (log45^{1-s}55^s - log2(0.5 + \delta_{12}) - 2(log45^{1-s}55^s - log2\delta_{12}) = 0$

$\Rightarrow log⁡(2.45^{1-s}55^s)/45^{1-s}55^s - log2\delta_{12}(1 + 2\delta_{12}) = 0$

$\Rightarrow \delta_{12}(1 + 2\delta_{12}) = 1.$

And

$(log\, 2.45^{1-s}55^s - log2\delta_{22}) + (log45^{1-s}55^s - log2(0.5 + \delta_{22}) - 2(log45^{1-s}55^s - log2\delta_{22}) = 0$

$\Rightarrow log⁡(2.45^{1-s}55^s)/45^{1-s}55^s - log2\delta_{22}(1 + 2\delta_{22}) = 0$

$\Rightarrow \delta_{22}(1 + 2\delta_{22}) = 1,$

From primal-dual relation

$\frac{10.16^{1-s}24^s}{Q_1} = \delta_{11}\sqrt{v(\delta, s)},$

$\frac{45^{1-s}55^s S_1{}^2}{Q_1} = \delta_{12}\sqrt{v(\delta, s)},$

$\frac{45^{1-s}55^s Q_1}{2} = \delta_{13}\sqrt{v(\delta, s)},$

$45^{1-s}55^s S_1 = \delta_{14}\sqrt{v(\delta, s)},$

$$\frac{15.8^{1-s}12^s}{Q_2} = \delta_{21}\sqrt{v(\delta,s)},$$

$$\frac{(115^{1-s}135^s+23^{1-s}27^s)S_2{}^2}{2Q_2} = \delta_{22}\sqrt{v(\delta,s)},$$

$$\frac{115^{1-s}135^sQ_2}{2} = \delta_{23}\sqrt{v(\delta,s)},$$

$$115^{1-s}135^sS_2 = \delta_{24}\sqrt{v(\delta,s)}.$$

Solutions of above relations with different values of weights are presented in table-10.4.

Table -10.4 (Optimal solution for FMSGP problem)

s	1 − s	Optimal dual variables	Optimal primal variables	Optimal values objectives	
				$v(\delta,s)$	$Tac(Q,S)$
0.1	0.9	$\delta_{11}{}^* = 0.5$, $\delta_{12}{}^* = 0.5$, $\delta_{13}{}^* = 1, \delta_{14}{}^* = 1$, $\delta_{21}{}^* = 0.5$, $\delta_{22}{}^* = 0.5$, $\delta_{23}{}^* = 1, \delta_{24}{}^* = 1$.	$S_1{}^* = 1.971$, $Q_1{}^* = 3.912$, $S_2{}^* = 0.774$, $Q_2{}^* = 1.549$.	8187.095	195.347
0.3	0.7	$\delta_{11}{}^* = 0.5$, $\delta_{12}{}^* = 0.5$, $\delta_{13}{}^* = 1, \delta_{14}{}^* = 1$, $\delta_{21}{}^* = 0.5$, $\delta_{22}{}^* = 0.5$, $\delta_{23}{}^* = 1, \delta_{24}{}^* = 1$.	$S_1{}^* = 2.008$, $Q_1{}^* = 4.015$, $S_2{}^* = 0.795$, $Q_2{}^* = 1.590$.	9205.062	206.989
0.5	0.5	$\delta_{11}{}^* = 0.5$, $\delta_{12}{}^* = 0.5$, $\delta_{13}{}^* = 1, \delta_{14}{}^* = 1$, $\delta_{21}{}^* = 0.5$, $\delta_{22}{}^* = 0.5$, $\delta_{23}{}^* = 1, \delta_{24}{}^* = 1$.	$S_1{}^* = 2.045$, $Q_1{}^* = 4.090$, $S_2{}^* = 0.816$, $Q_2{}^* = 1.633$.	10349.600	219.326
0.7	0.3	$\delta_{11}{}^* = 0.5$, $\delta_{12}{}^* = 0.5$, $\delta_{13}{}^* = 1, \delta_{14}{}^* = 1$, $\delta_{21}{}^* = 0.5$, $\delta_{22}{}^* = 0.5$, $\delta_{23}{}^* = 1, \delta_{24}{}^* = 1$.	$S_1{}^* = 2.083$, $Q_1{}^* = 4.166$, $S_2{}^* = 0.838$, $Q_2{}^* = 1.677$.	11636.450	232.372
0.9	0.1	$\delta_{11}{}^* = 0.5$, $\delta_{12}{}^* = 0.5$, $\delta_{13}{}^* = 1, \delta_{14}{}^* = 1$, $\delta_{21}{}^* = 0.5$, $\delta_{22}{}^* = 0.5$, $\delta_{23}{}^* = 1, \delta_{24}{}^* = 1$.	$S_1{}^* = 2.122$, $Q_1{}^* = 4.244$, $S_2{}^* = 0.861$, $Q_2{}^* = 1.722$.	13083.300	246.191

10.5   Conclusion

In this chapter, a fuzzy EOQ model with shortages under fully backlogging and constant demand is formulated and solved by fuzzy signomial geometric programming (FSGP) technique. Here coefficients are expressed as triangular fuzzy number (TFN). In future other types of fuzzy numbers would be used. The methodology proposed in this chapter may also be applicable to other EOQ models. Our approach provide here a simple EOQ model, but in the future it should be used many complex EOQ models. For future research of uncertainty in economic order quantity (EOQ) model, by using different type of fuzzy numbers such as pentagonal, hexagonal fuzzy numbers of generalized fuzzy numbers be analytically more challenging and interesting. Inflation plays an important role in present day-to-day life, but we have neglected it. Therefore, consideration of inflation problem would be more realistic.

Chapter 11

Conclusions & Future Work

The study was developed some inventory model in crisp and fuzzy environment. The reason for choosing this subject was that inventory model is one of the most important tasks in production department. Selecting the right inventory can decreases total average cost and maximizes profit. All models primarily focus on minimizes total average cost and maximizes profit. *In real life situation uncertainty makes an inventory model more interesting.* ***This chapter gives an overview of the complete research project and its conclusions and outlines further directions to extend and advance the current research.***

11.1 Conclusions

The following achievements and conclusion of this research work:

a) **In chapter 3**, an inventory model with cost of interest time depended holding cost without shortages is formulated and solved. Here total cost of interest is inversely related to set-up cost, production quantity and directly related to cycle of length.

Differencing from other inventory models published in the literature is that, the inventory model is solved by various techniques in crisp and fuzzy environment. Here the model is solved by geometric programming (GP), non-linear programming (NLP), fuzzy geometric programming (FGP) and fuzzy non-linear programming (FNLP) technique respectively and we have seen that fuzzy geometric programming (FGP) gives better result than any other solution techniques.

b) **In chapter 4**, an inventory model with three different cases has been developed and solved in a crisp and fuzzy environment. In fuzzy environment considered a recent idea that is nearest interval approximation (NIA) method and parametric interval-valued function.

We have developed the inventory system, in based on three different cases. Here we have considered a special condition that the demand falls to zero in a time interval $(t_0 \leq t \leq t_e)$ for an unexpected situation (flood, strike, earthquake, etc.). In case 1, demand falls to zero before start the deterioration, in case 2, demand falls to zero after start the deterioration and in case 3, no unexpected condition account here.

c) **In chapter 5**, an inventory model with time depended demand under a space constraints has been developed and solved in a crisp and fuzzy environment. Geometric programming (GP) provides a powerful tool for solving a variety of imprecise optimization problem. Here the problem is solved using geometric programming (GP) technique.

Differencing from other inventory models published in the literature is that, the development of the inventory model is based on unit production cost, which is continuous function of demand and set-up cost. Also we have considered the holding cost as a time depended.

d) **In chapter 6**, an economic order quantity (EOQ) model with unit production cost, time depended holding cost, without shortages is formulated and solved.

Here, the inventory model is solved by Max-Min FGP and parametric FGP technique respectively and compared the numerical results.

e) **In chapter 7**, an inventory model has been developed and solved in crisp and fuzzy environment. In fuzzy, considered all cost functions in the type of triangular fuzzy number (TFN) and solved using fuzzy grade-mean integration representation technique.

Differencing from other inventory models published in the literature is that, the development of the inventory system is based on the time depended demand. During $0 \leq t \leq t_w$ the demand is increasing as $a + bt$, and in $t_w \leq t \leq t_1$ the demand is decreasing as $a - bt$. Shortage is also considered here.

f) **In chapter 8**, we have developed a fuzzy inventory model with constant demand, without shortages using possibilistic approach. In this chapter we have developed the concept of possibility theory and possibilistic moment generating function and some statistical concept as mean, variance, standard deviation on this economic order quantity (EOQ) model.

Differencing from other inventory models published in the literature is that, here a new idea that is Bell-Shaped fuzzy membership function (MF) is used and developed the various type of moment generating functions.

g) **In chapter 9**, we have proposed a fuzzy inventory model with constant demand and with shortages under fully backlogging using possibilistic approach. Three type of possibilistic mean values as lower possibilistic$(E_L(A))$, upper possibilistic $(E_R(A))$ and crisp possibilistic (E(A)) of total average cost function developed here.

First time Apadoo et al. (2010) developed same type of inventory model, using fuzzy Gaussian membership function, but in this thesis, developed the model by using a new idea that is Bell-Shaped fuzzy membership function.

h) **In chapter 10**, a fuzzy economic order quantity (EOQ) model with shortages under fully backlogging and constant demand is formulated and solved. Here the model is solved by fuzzy signomial geometric programming (FSGP) technique. Fuzzy signomial geometric programming (FSGP) technique provides a powerful technique for solving some special non-linear problems.

Here we have proposed a new idea that is fuzzy modified signomial geometric programming (FMSGP) and some necessary theorems have been derived.

11.2　Future Work

The research work presented in this book has some limitations which should be overcome;

- In this book mainly inventory model under single item is developed. But in future, the multi-item inventory model needs to develop.

- All models developed here under the consideration of negligible lead time, but lead time plays an important role in an inventory system. So in future developed an inventory model under the consideration of lead time will be more challenging and interesting.

- In fuzzy environment, here we have considered all cost coefficients in a triangular fuzzy number (TFN), but in future consideration other type of fuzzy number that is, trapezoidal fuzzy number (TrFN), pentagonal fuzzy number (PFN), hexagonal fuzzy number (HFN), etc. may be consider for the inventory model.

- In this book some models are developed without shortages and some models are developed with shortages under fully backlogging but we have not considered shortages with partially backlogged. So consideration of shortages with partially backlogging should be more challenging.

- Inflation plays an important role in an inventory management (IM) but it neglect here. So consideration of inflation in an inventory more realistic.

- Inventory models are developed in this thesis under deterministic approach but probabilistic approach is also more interesting.

- In this book only cost parameter considers as a fuzzy parameters. In future, other inventory related parameters (e.g. lead time, production capacity, deterioration rate etc.) may be considered as fuzzy parameters.

- Finally it would be account to external factors, especially the human factor in more detail but this left outside the scope of this research work. But human behavior can certainly influence the measures of the inventory model in crucial way. Finding a way to standardize this influence would make comparison between organizations more reliable and provides even more insight.

BIBLIOGRAPHY

A. Charnes, W.W. Copper, B. Golany and J. Masrers, "Optimal dwsine modification by geometric programming and constrained stochastic network models" International Journal of System Science 19 (1988) 825-844.

Alamri, A.A., Balkhi, Z.T. (2007) "The Effects of Learning and For getting on the Optimal Production Lot size for Deteriorating Items with Time Varying Demand and Deterioration Rates", International Journal of Production Economics, 107(1): 125-138.

Ali Ekici. (2013) "An Improved Model for Supplier Selection under Capacity Constraint and Multiple Criteria", International Journal of Production Economics, 141: 574-581.

Apadoo, S.S., Gajpal, Y. & Bhatt, S. K. "A Gaussian Fuzzy Inventory EOQ Model Subject to Inaccuracies In Model Parameters", International Conference on Business and Management Mumbai – India December 26th 2014.

Appadoo, S. S., Bhatt, S. K., Bector, C. R., "Application of possibility theory to investment decisions", FuzzyOptimization and Decision Making, 7, 2008, 35-57.

Appadoo, S.S, Bector, C.R, Bhatt, S.K (2012) "Fuzzy EOQ model using possibilistic approach",Journalof Advances in Management Research, Vol. 9 Iss: 1, pp.139 – 164.

Arora, V., Chan, F.T.S and Tiwari, M.K. (2010) "An Integrated Approach for Logistic and Vendor Managed Inventory in Supply Chain", Expert Systems with Applications, 8, 8.doi: 10: 1016/j.eswa.2009.05.016.

Barbarosoglu, G. (2000) "An Integrated Supplier-Buyer Model for Improving Supply Chain Coordination", Production Planning and Control, 11(8) : 732-741.

Baumol, W.J., Vinod, H.D. (1970) "An Inventory Theoretic Model of Freight Transport Demand", Management Science, 16 : 413-421. Beightler.C. and D.Philips. (1976) Applied Geometric Programming, John Wiley and Sons, New York.

Bellman, R.E., Zadeh, L.A. (1970) "Decision Making in a Fuzzy Environment", Management Science, 17: B141-B164.

Ben-daya, M., Hariga, M. (2003) "Lead Time Reduction in a Stochastic Inventory System with Learning Consideration", International Journal of Production Research, 41(3) : 571-579.

Bjork, K.M. (2009) "An Analytical Solution to a Fuzzy Economic Order Quantity Problem". International Journal of Approximate Reasoning, 50 : 485-493.

Borgonovo, E., Peccati, L. (2009) "Financial Management in Inventory Problems : Risk Averse Vs Risk Neutral Policies", International Journal of Production Economics, 118 : 233-242.

Buffa, F.P., Reynolds, J.I. (1977) "The Inventory Transport Model with Sensitivity Analysis by Indifference Curves", Transportation Journal, 17 : 83-90.

Bylka, S. (2003) "Competitive and Cooperative Policies for the Vendor- Buyer System", International Journal of Production Economics, 81-82 : 533-544.

Carlsson, C. and Fuller, R., "On Possibilistic Mean Value and Variance of Fuzzy Numbers", Fuzzy Sets and Systems 122, 2001, 315- 326.

Carlsson, C. and Fuller, R. (2002) "Fuzzy Reasoning in Decision Making and Optimization", Physics-Verlag.

Chan, W.M., Ibrahim, R.N., Lochert, P.B. (2003) "A New EPQ Model: Integrating Lower Pricing, Rework and Reject Situations", Production, Planning and Control : The Management of Operations, 12 : 588-595.

Chang, C.T., Ouyang, L.Y and Teng, J.T. (2003) "An EOQ model for Deteriorating Items Under Supplier Credits Linked to Ordering Quantity", Applied Mathematical Modelling, 27 : 983-996.

Chang, H.C. (2004) "An Application of Fuzzy Set Theory to the EOQ Model with Imperfect Quality Items", Computers and Operations Research, 31 : 2079-2092.

Chen, L.H & Kang, F.S. (2007) "Integrated Vendor Buyer Cooperative Inventory Models with Variant Permissible Delay in Payments", European Journal of Operational Research, 183 : 658-673.

Chen, S.H and Hsieh, C.H. (1998) "Graded Mean Integration Representation of Generalized Fuzzy Numbers", In Proceedings of the Sixth Conference on Fuzzy Theory and Its Applications, Chinese Fuzzy Systems Association, Taiwan, pp.1-6.

Chen, S.H., Hsieh, C.H. (2000) "Representation, Ranking, Distance and Similarity of L-R Type Fuzzy Number and Application", Australian Journal of Intelligent Information Processing Systems, 6 : 217-299.

Chih Hsun Hsieh. (2002) "Optimization of Fuzzy Production Inventory Models", Information Sciences, 146 : 29-40.

Chi Kin Chan, Lee, Y.C.E and Goyal, S.K. (2010) "A Delayed Payment Method in Coordinating a Single-Vendor Multi-Buyer Supply Chain", International Journal of Production Economics, 127(1) : 95-102.

Chiu, H.N., Chen, H.M. (2005) "An Optimal Algorithm for Solving the Dynamic Lot sizing Model with Learning and Forgetting in Setups and Production", International Journal of Production Economics, 95(2) : 179-193.

Clark, A.J (1992). An informal survey of multi-echelon inventory theory, Naval Research Logistics Quarterly, 19, 621-650.

D. Dubois and H. Prade (1987) "The mean value of a fuzzy number", Fuzzy Sets and Systems, 24:279-300.

Dickson, G.W. (1966) "An Analysis of Vendor Selection Systems and Decisions", Journal of Purchasing, 2(1) : 5-17.

Dutta, D & Kumar, P (2012) "Fuzzy inventory without shortages using trapezoidal fuzzy number with sensitivity analysis", IOSR Journal of Mathematics, Vol. 4(3), 32-37.

Dutta, D, Rao, J.R., & Tiwary R.N (1993). "Effect of tolerance in fuzzy linear fractional programming", Fuzzy Sets and Systems, 55, 133-142.

Duffin, R.J, & E.L. Peterson, "Duality Theory for Geometric Programming", SIAM Jour. Of App. Math., 14(1966), 1307-1349.

Duffin, R.J., E.L. Peterson & C. Zener (1967)" Geometric Programming-Theory and Applications", John Wiley, New York, 1967.

Ecker, J. G., Kupferschmid, M., andSacher, R. S. (1984) "Comparison of a Special-Purpose Algorithm with General-Purpose Algorithms for Solving Geometric Programming Problems", Journal of Optimization Theory and Applications, Vol. 43, pp. 237–262.

Eroglu, A., Ozdemir, G. (2007) "An Economic Order Quantity Model with Defective Items and Shortages", International Journal of Production Economics, 106 : 544-549.

Ghodsypour, S.H., O'Brien, C. (2001) "The Total Cost of Logistics in Supplier Selection, under Conditions of Multiple Sourcing, Multiple Criteria and Capacity Constraint", International Journal of Production Economics, 73(1) : 15-27.

Gour Chandra Mahata and Adrijit Goswami (2013) "Fuzzy Inventory Models for Items with Imperfect Quality and Shortage Backordering under Crisp and Fuzzy Decision Variables", Computers and Industrial Engineering, 64 : 190-199.

Goyal, S.K. (1985) "EOQ under Conditions of Permissible Delay in Payments", Journal of Operational Research Society, 36 : 335-338.

Gupta, O.K. (1992) "A Lot Size Model with Discrete Transportation Costs", Computers and Industrial Engineering, 22 : 397-402.

Hadley, G., Whitin, T.H. (1963) "Analysis of Inventory Systems", Prentice Hall, Englewood Cliffs, New Jersey.

Hamacher, Leberling, H & Zimmermann, H.J (1978)" Sensitivity Analysis in fuzzy linear Programming", Fuzzy Sets and Systems, 1, 269-281.

Hariga, M.A. (1996) "Optimal EOQ models for deteriorating items with time-varying demand", J. Oper. Res. Soc. 47, 1228–1246 .

Harris, Ford W. (1990). "How Many Parts to Make at Once". Operations Research. 38(6): 947.

Ho, W., Xu, X, Dey, P.K. (2010) "Multi Criteria Decision Making Approaches for Supplier Evaluation and Selection : A Literature Review", European Journal of Operational Research, 202(1) : 16-24.

Huang, Y. (2007) "Economic Order Quantity under Conditionally Permissible Delay in Payments", European Journal of Operational Research, 176(2) : 911-924.

Islam, S. & RoyT.K (2006) "A fuzzy EPQ model with flexibility and reliability consideration and demand depended unit Production cost under a space constraint: A fuzzy geometric programming approach", Applied Mathematics and Computation, 176(2), 531-544.

Islam, S. & Roy, T.K. (2010) "Multi-Objective Geometric-Programming Problem and its Application", Yugoslav Journal of Operations Research, 20, 213-227.

Jaber, M.Y., Goyal, S.K., Imran, M. (2008) "Economic Production Quantity Model for Items with Imperfect Quality Subject to Learning Effects", International Journal of Production Economics, 115 : 143-150.

Jaber, M.Y. (2011) "Learning Curves Theory Models and Applications", CRC Press (Tailor and Francis Group), FL, Baco, Raton.

Jaber, M.Y., Bonney, M. (2007) "Economic Manufacture Quantity (EMQ) Model with Lot Size Dependent Learning and Forgetting Rates", International Journal of Production Economics, 108(1-2), 359-367.

Jaber, M.Y., Guiffrida, A.L. (2007) "Observations on the Economic Manufacture Quantity Model with Learning and Forgetting", International Transactions in Operational Research, 14(2) : 91-104.

Jaggi, C.K., Pareek, S., Sharma, A. (2012) "Nidhi, Fuzzy Inventory Model for Deteriorating Items with Time varying Demand and Shortages", American Journal of Operational Research. 2(6), 81–92.

Johnson, L.A., Montgomery, D.C. (1974) "Operations Research in Production Planning", Scheduling and Inventory Control, Wiley New York.

Johnson, M.R., Wang, M.H. (1998) "Economical Evaluation of Disassembly Operations for Recycling, Remanufacturing and Reuse", International Journal of Production Research, 36(12): 3227-3252.

Kao, C and Hsu, W.K. (2002) "Lot size Reorder Point Inventory Model with Fuzzy Demand", Computer and Mathematic with Application, 43: 1291-1302.

K. A. M. Kotb and H. A. Fergany (2011) "Multi-Item EOQ Model with Varying Holding cost: A Geometric Programming Approach," International Mathematical Forum, Vol. 6, No. 23, pp. 1135-1144.

Karwowski, W and Evans, R.W. (1986) "Fuzzy Concepts in Production Management Research : A Review", International Journal of Production Research, 24 : 129-147.

Katagiri, H., Ishii, H. (2002) "Fuzzy Inventory Problems for Perishable Commodities", European Journal of Operational Research, 138: 545-553.

Kaufmann, A. (1975) "Introduction to the Theory of Fuzzy Subsets", Vol.1, Academic Press, New York.

Kiesmuller, G.P., Minner, S., Kleber, R. (2004) "Managing Dynamic Product Recovery : An Optimal Control Perspective", pp.221-247.

Kim, B. (2000) "Coordinating an Innovation in Supply Chain Management", European Journal of Operational Research, 123 : 568-584.

Kochenberger, G.A. (1971) "Inventory Models: Optimization by Geometric Programming", Decision Sciences, 2 : 193-205.

Konstantaras, I., Skouri, K and Jaber, M.Y. (2012) "Inventory Models for Imperfect Quality Items with Shortages and Learning in Inspection", Applied Mathematical Modeling, 36: 5334-5343.

Khun, H.W & Tucker A.W (1951). "Non-linear programming, proceeding second Berkeley symposium Mathematical Statistic and probability (ed) Nyman" „J. University of California press 481-492.

Kotb A.M , Hlaa.A. & Fergancy (2011). "Multi-item EOQ model with both demand-depended unit constant varying Lead time via Geometric Programming", Applied Mathematics, 2011, 2, 551-555.

Kun. Jen Chung, Leopoldo Eduardo Cardenas – Barron. (2012) "The Complete Solution Procedure for the EOQ and EPQ Inventory Models with Linear and Fixed Backorder Costs", Mathematical and Computer Modeling, 55 : 2151-2156.

Liang, Y. & Zhou, F. (2011), "A two warehouse inventory model for deteriorating items under conditionally permissible delay in Payment", Applied Mathematical Modeling, 35, 2221-2231

Liu, S. T. (2006). "Posynomial Geometric-Programming with interval exponents and coefficients", European Journal of Operations Research,168(2), 345-353

Maddah, B., Jaber, M.Y. (2008) "Economic Order Quantity for Items with Imperfect Quality", Revisited, International Journal of Production Economics, 112 : 808-815.

Mandal, W.A. & S.Islam (2016). "Fuzzy Inventory Model for Deteriorating Items, with Time Depended Demand, Shortages, and Fully Backlogging", Pak. j. stat. oper.res. Vol. .XII No.1 2016 pp101-109.

Mandal, W.A. & S.Islam (2015). "Fuzzy Inventory Model for Power Demand Pattern with Shortages, Inflation Under Permissible Delay in Payment", International Journal of Inventive Engineering and Sciences,3(8).

Mandal, W.A. & S.Islam (2015). "Fuzzy Inventory Model for Weibull Deteriorating Items, with Time Depended Demand, Shortages, and Partially Backlogging", International Journal of Engineering and Advanced Technology, 4(5).

Mandal, W.A. & S.Islam (2016). 'A Fuzzy E.O.Q Model with Cost of Interest, Time Depended Holding Cost, With-out Shortages under a Space Constraint: A Fuzzy Geometric Programming and Non-Linear Programming Approach", International Journal of Research on Social and Natural Sciences Vol. I Issue 1 June 2016, 134-147.

Maity, M.K. (2008). "Fuzzy inventory model with two ware house under possibility measure in fuzzy goal", European Journal Operation Research, 188,746-774

Mahata, G.C and Mahata, P. (2011) "Analysis of a Fuzzy Economic Order Quantity Model for Deteriorating Items under Retailer Partial Trade Credit Financing in a Supply Chain", Mathematical and Computer Modelling, 53 : 1621-1636.

Mirko Vujosevic, Dobrila Petrovic, Radivoj Petrovic (1996) "EOQ Formula when inventory cost is fuzzy" International Journal Production Economics, vol. 45, pp. 499-504.

Mishra, P., Shah, N.H. (2008) "Inventory management of time dependent deteriorating items with salvage value", Applied Mathematical Sciences, 2 (16), 793–798.

Mondal, S and Maiti, M. (2003) "Multi Item Fuzzy EOQ Models using Genetic Algorithm", Computers and Industrial Engineering, 44 : 105-117.

Mohamad Y. Jaber, Ahmed, M.A., El Saadany. (2011) "An Economic Production and Remanufacturing Model with Learning Effects", International Journal of Production Economics, 131: 115-127.

Moon, I., Giri, B.C. Ko, B. (2005) "Economic order quantity models for ameliorating/deteriorating items under inflation and time discounting", Eur. J. oper. Res., 162,773–785

Naddor, E. (1966) "Inventory Systems", John Wiley & Sons, New York.

N. K. Mandal, T. K. Roy and M. Maiti (2006) "Inventory Model of Deteriorated Items with a Constraints: A Geometric Programming Approach," European Journal of Operational Research, Vol. 173, No. 1, pp. 199-210. doi:10.1016/j.ejor.2004.12.002

Ouyang, L.Y., Chang, H.C. (2002) "A Mini-Max Distribution Free Procedure for Mixed Inventory Models Involving Variable Lead Time with Fuzzy Lost Sales", International Journal of Production Economics,
76(1) : 1-12.

Papachristos, S., Konstantaras, I. (2006) "Economic Ordering Quantity Models for Items with Imperfect Quality", International Journal of Production Economics, 100 : 148-154.

Park, K.S. (1987) "Fuzzy Set Theoretic Interpretation of Economic Order Quantity". IEEE Transactions on Systems, Man and Cybernatics, SMC 17: 1082-1084.

Paseka,A., Appadoo, S.S., Thavaneswaran, A., "Possibilistic moment generating functions", Applied Mathematics Letters Volume 24, Issue 5, May 2011, Pages 630635.

Passy, U. and D. J. Wilde, "Generalized Polynomial Optimization" SIAM Jour. Appli. Math., 15(1967), 1344-1356.

Passy, U. and D. J. Wilde, "A Geometric Programming Algorithm for Solving Chemical Equilibrium Problems" SIAM Jour. Appli. Math., 16(1968), 363-373.

R. J. Duffi, E. L. Peterson, C. M. Zener, "Geometric programming- theory and applications", Wiley, New York, (1967).

R Uthaykumar, M Valliathal (2011) "Fuzzy economic production quantity model for weibull deteriorating items with ramp type of demand", International Journal of Strategic Decision Sciences, vol. 2, no. 3, pp. 55-90.

Rosenblatt, M.J and Lee, H.L. (1986) "Economic Production Cycles with Imperfect Production Processes", IIE Transactions, 18: 48-55.

Roy.T.K, Maiti, M. (1997), "A Fuzzy EOQ Model with Demand –Dependent Unit Cost under Limited Storage Capacity", European Journal of Operational Research 99, 425-432.

Roy, T.K. & Mayty, M. (1995). :A fuzzy inventory model with constraints", Opsearch, 32(4) (1995) 287-298.

S. T. Liu (2008) "Posynomial geometric programming with interval exponents and Coefficients", European Journal of Operational Research, 186 (1) (2008) 17-27.

S. Dey, T. K. Roy (2015), "Optimum shape design of structural model with imprecise coefficient by parametric geometric programming", Decision Science Letters, 4 (2015) 407-418.

S. Islam, T. K. Roy, "Modified Geometric programming problem and its applications", J. Appt. Math and computing, 17 (1-2) (2005) 121-144.

S. Islam, T. K. Roy, "A fuzzy EPQ model with flexibility and reliability consideration and demand depended unit Production cost under a space constraint: A fuzzy geometric programming approach", Applied Mathematics and Computation, 176 (2) (2006) 531-544.

S. Islam, T. K. Roy, "Multi-Objective Geometric-Programming Problem and its Application", Yugoslav Journal of Operations Research, 20 (2010) 213-227.

Salameh, M.K., Jaber, M.Y. (2000) "Economic Production Quantity Model for Items with Imperfect Quality", International Journal of Production Economics, 64 : 59-64.

Sarker, B.R., Mukherjee, S., Balan, C.V. (1997) "An order-level lot size inventory model with inventory-level dependent demand and deterioration", Int. J. Prod. Eco. 48,227–236.

Schrady, D.A. (1967) "A Deterministic Model for Repairable Items", Naval Research Logistics Quarterly 14(3) : 391-398.

Shen, S.Y and Honda, M. (2009) "Incorporating Lateral Transfers of Vehicles and Inventory into an Integrated Replenishment and Routing Plan for a Three Echelon Supply Chain", Computers and Industrial Engineering, 56 : 754-775.

Silver, E.A., Pyke, D.F., Peterson, R. (1998) "Inventory Management and Production Planning and Scheduling", John Wiley and Sons, New York.

Sphicas, G.P. (2006) "EOQ and EPQ with Linear and Fixed Backorder Costs, Two Cases Identified and Models Analysed without Calculus", International Journal of Production Economics, 100 : 59-64.

Swenseth, S.R., Godfrey, M.R. (2002) "Incorporating Transportation Costs into Inventory Replenishment Decisions", International Journal of Production Economics, 77 : 113-130.

Teng, J., Chang, C and Goyal, S.K. (2005) "Optimal Pricing and Ordering Policy under Permissible Delay in Payments", International Journal of Production Economics, 97(2) : 121-129.

Teunter, R.H. (2001) "Economic Ordering Quantities for Recoverable Item Inventory Systems", Naval Research Logistics, 48(6) : 484-495.

Van Houtum, G.J., Inderfurth, K and Zijm, W.H.M. (1996) "Materials Coordination in Stochastic Multi-Echelon Systems", European Journal of Operational Research, 95 : 1-23.

Vijayan, T and Kumaran, M. (2008) "Inventory Models with a Mixture of Backorders and Lost Sales under Fuzzy Cost", European Journal of Operational Research, 189: 105-119.

Vujosevic, M., Petrovic, D., Petrovic, R. (1996) "EOQ Formula when Inventory Cost is Fuzzy", International Journal of Production Economics, 45 : 499-504.

Wagner, S.M. (2006) "A Firm's Responses to Deficient Suppliers and Competitive Advantage", Journal of Business Research, 59 : 686-695. Weber, C.A., Current, J.R., Benton, W.C. (1991) Vendor Selection Criteria and Methods, European Journal of Operational Research, 50(1): 2-18.

Wee, H.A (1995) "deterministic lot-size inventory model for deteriorating items with short-ages and a declining market", Comp. Oper. Res. 22, 345–356.

Wee, H.M., Jonas Y and Chen, M.C. (2007) "Optimal Inventory Model for Items with Imperfect Quality and Shortage Backordering", Omega, 5: 7-11.

Xu, K., Dong, Y and Evers, P.T. (2001) "Towards Better Coordination of the Supply Chain", Transportation Research Part-E : Logistics and Transportation Review, 37 : 35-54.

Yager. R.R.(1981), "A Procedure for Ordering Fuzzy Subsets of the Unit Interval", Information Sciences 24, pp.143-161

Yao, J.S and Chiang, J. (2003) "Inventory without Backorder with Fuzzy Total Cost and Fuzzy Storing Cost Defuzzified by Centroid and Signed Distance", European Journal of Operational Research, 148: 401-409.

Yao, J.S., Chang, S.C., Su, J.S. (2000) "Fuzzy Inventory Without Backorder for Fuzzy Order Quantity and Fuzzy Total Demand Quantity", Computers and Operations Research, 27 : 935-962.

Zadeh, L.A. (1965) "Fuzzy Sets", Information and Control, 8 : 338-353.

Zadeh, L.A. (1978) "Fuzzy Sets as a Basis for a Theory of Possibility", Fuzzy Sets and Systems, I(1) : 3-28.

Zhao, Q.H., Wang, S.Y., Lai, K., Xia, G.P. (2004) "Model and Algorithm of an Inventory Problem with the Consideration of Transportation Cost", Computers and Industrial Engineering, 46 : 389-397.

Zimmer, K. (2002) "Supply Chain Coordination with Uncertain JIT Delivery", International Journal of Production Economics, 77: 1-15.

Zimmermann, H.J. (1983) "Using Fuzzy Sets in Operational Research", European Journal of Operational Research, 13: 201-206.

Zimmermann, H.J. (1991) "Fuzzy Set Theory and its Applications", Second Edition, Kluwer – Nijhoff, Boston.

Zimmermann, H.J. (2001) "Fuzzy Set Theory and its Applications", Fourth Edition, Kluwer Academic, Boston.